student study
ART NOTEBOOK

FOUNDATIONS IN
MICROBIOLOGY

Second Edition

KATHLEEN TALARO
Pasadena City College

ARTHUR TALARO
Pasadena City College

WCB Wm. C. Brown Publishers

Dubuque, IA   Bogotá   Buenos Aires   Caracas   Chicago   Guilford, CT   London
Madrid   Mexico City   Seoul   Singapore   Sydney   Taipei   Tokyo   Toronto

A Times Mirror Company

The credits section for this book begins on page 157 and
is considered an extension of the copyright page.

A Times Mirror Company

ISBN 0–697–28329-1

Printed in the United States of America by Wm. C. Brown Communications, Inc.,
2460 Kerper Boulevard, Dubuque, IA 52001

10   9   8   7   6   5   4   3   2   1

TO INSTRUCTORS AND STUDENTS

The Student Study Art Notebook is designed to help in your study of microbiology. The notebook contains art taken from the text and overhead transparencies; thus you can take notes during lectures, or jot down comments as you are reading through the chapters.

We hope this notebook, used along with your text, helps to make the study of microbiology easier for you.

DIRECTORY OF NOTEBOOK FIGURES

TO ACCOMPANY TALARO-TALARO
FOUNDATIONS IN MICROBIOLOGY, 2/E

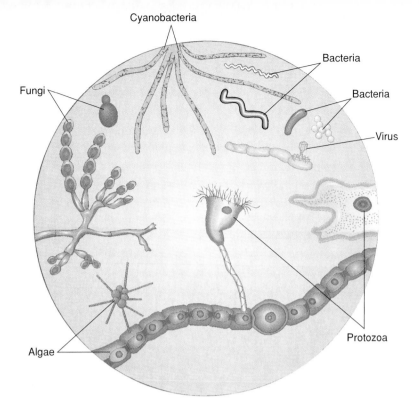

Diversity of the Microbial World
Figure 1.1

(a)

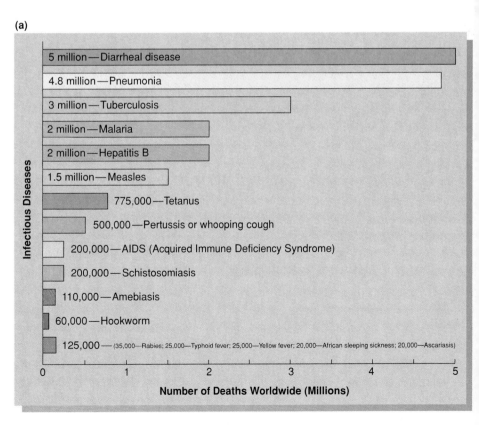

Infectious Diseases	Number of Deaths Worldwide (Millions)
5 million — Diarrheal disease	
4.8 million — Pneumonia	
3 million — Tuberculosis	
2 million — Malaria	
2 million — Hepatitis B	
1.5 million — Measles	
775,000 — Tetanus	
500,000 — Pertussis or whooping cough	
200,000 — AIDS (Acquired Immune Deficiency Syndrome)	
200,000 — Schistosomiasis	
110,000 — Amebiasis	
60,000 — Hookworm	
125,000 — (35,000—Rabies; 25,000—Typhoid fever; 25,000—Yellow fever; 20,000—African sleeping sickness; 20,000—Ascariasis)	

Worldwide Mortality from Infectious Diseases
Figure 1.3a

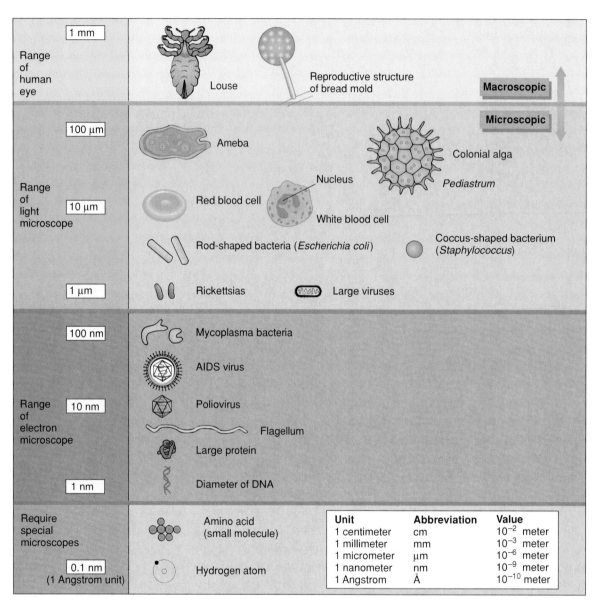

Range of human eye	1 mm	
		Louse
		Reproductive structure of bread mold
		Macroscopic
		Microscopic
Range of light microscope	100 μm	Ameba
		Colonial alga *Pediastrum*
		Nucleus
	10 μm	Red blood cell
		White blood cell
		Rod-shaped bacteria (*Escherichia coli*)
		Coccus-shaped bacterium (*Staphylococcus*)
	1 μm	Rickettsias Large viruses
Range of electron microscope	100 nm	Mycoplasma bacteria
		AIDS virus
	10 nm	Poliovirus
		Flagellum
		Large protein
	1 nm	Diameter of DNA
Require special microscopes		Amino acid (small molecule)
	0.1 nm (1 Angstrom unit)	Hydrogen atom

Unit	Abbreviation	Value
1 centimeter	cm	10^{-2} meter
1 millimeter	mm	10^{-3} meter
1 micrometer	μm	10^{-6} meter
1 nanometer	nm	10^{-9} meter
1 Angstrom	Å	10^{-10} meter

Common Measurements Encountered in Microbiology
Figure 1.5

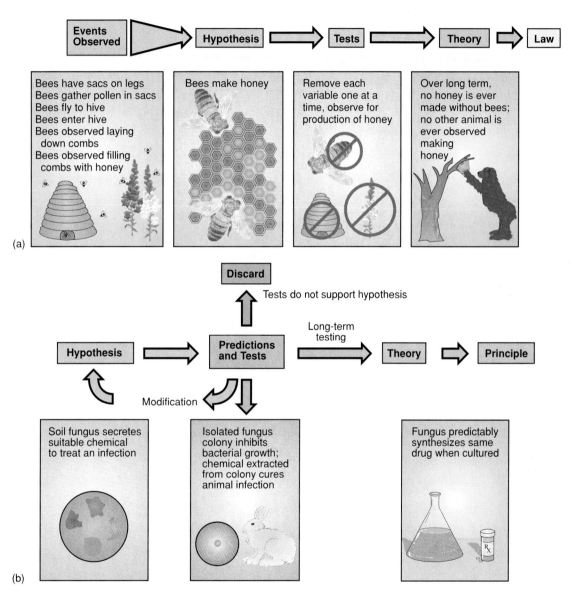

Inductive and Deductive Approaches to the Scientific Method
Figure 1.8

Subject: Testing the factors responsible for dental caries

Hypothesis: Dental caries (cavities) involve dietary sugar or microbial action or both.

Variables:

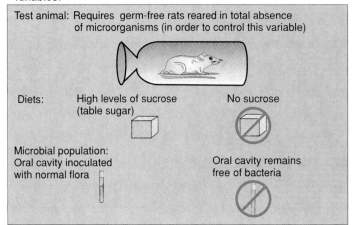

Test animal: Requires germ-free rats reared in total absence of microorganisms (in order to control this variable)

Diets: High levels of sucrose (table sugar)   No sucrose

Microbial population: Oral cavity inoculated with normal flora   Oral cavity remains free of bacteria

Experimental Protocol:

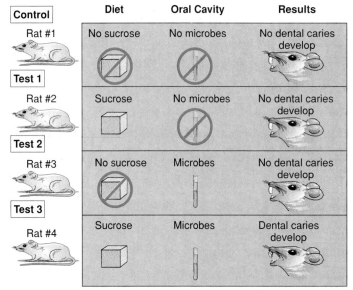

Control	Diet	Oral Cavity	Results
Rat #1	No sucrose	No microbes	No dental caries develop
Test 1			
Rat #2	Sucrose	No microbes	No dental caries develop
Test 2			
Rat #3	No sucrose	Microbes	No dental caries develop
Test 3			
Rat #4	Sucrose	Microbes	Dental caries develop

Conclusion: Dental caries will not develop unless both sucrose and microbial action are present. What other variables were not controlled?

Use of Controls in Experiments
Figure 1.9

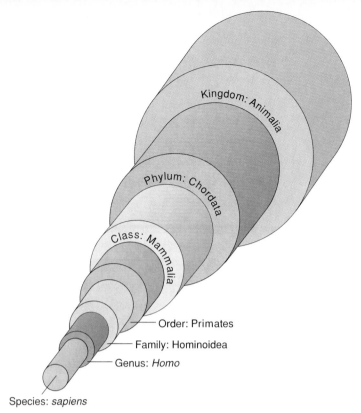

Levels in Classification
Figure 1.12

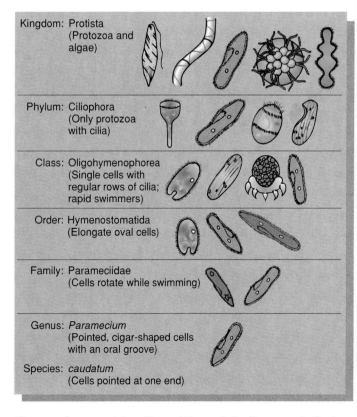

Paramecium caudatum **Traced through its Taxonomic Series**
Figure 1.13

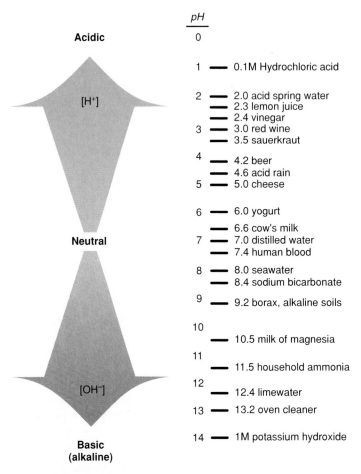

	pH
Acidic	0
[H⁺]	1 — 0.1M Hydrochloric acid
	2 — 2.0 acid spring water
	— 2.3 lemon juice
	— 2.4 vinegar
	3 — 3.0 red wine
	— 3.5 sauerkraut
	4 — 4.2 beer
	— 4.6 acid rain
	5 — 5.0 cheese
	6 — 6.0 yogurt
	— 6.6 cow's milk
Neutral	7 — 7.0 distilled water
	— 7.4 human blood
	8 — 8.0 seawater
	— 8.4 sodium bicarbonate
	9 — 9.2 borax, alkaline soils
	10
	— 10.5 milk of magnesia
	11
	— 11.5 household ammonia
	12
	— 12.4 limewater
	13 — 13.2 oven cleaner
[OH⁻]	14 — 1M potassium hydroxide
Basic (alkaline)	

The pH Scale for Various Substances and Habitats
Figure 2.12

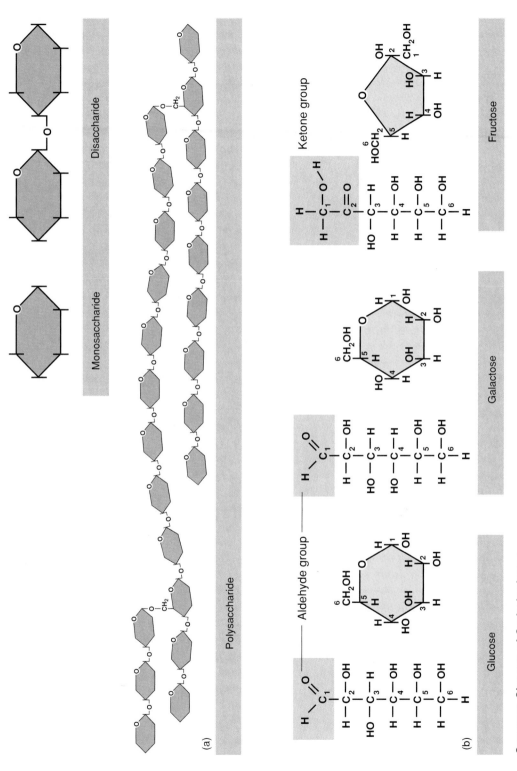

Common Classes of Carbohydrates
Figure 2.14

7

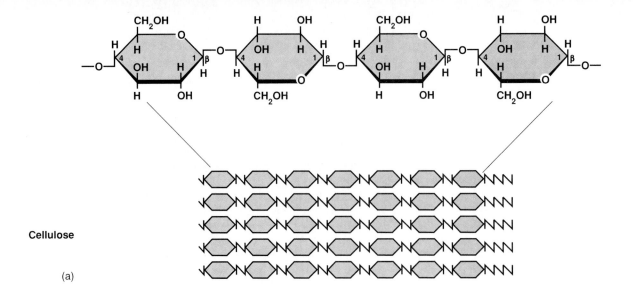

Cellulose

(a)

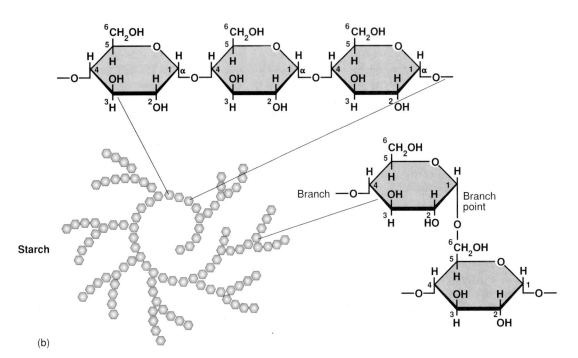

Starch

(b)

Polysaccharides
Figure 2.16

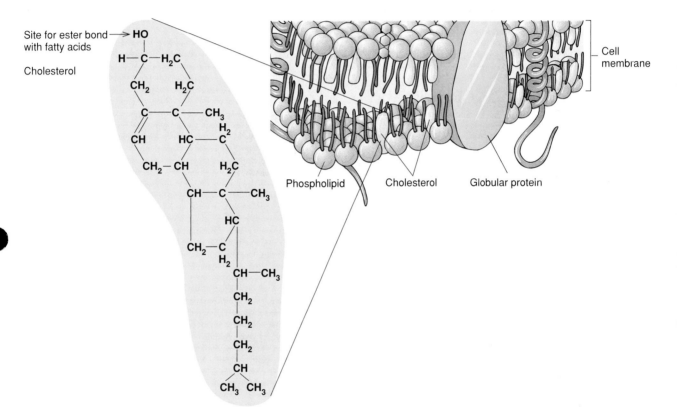

Site for ester bond → HO
with fatty acids

Cholesterol

H—C—H₂C

CH₂   H₂C

C—C—CH₃

CH   HC   H₂C

CH₂—CH   H₂C

CH—C—CH₃

HC

CH₂—C
        H₂

CH—CH₃

CH₂

CH₂

CH₂

CH

CH₃ CH₃

Phospholipid   Cholesterol   Globular protein

Cell
membrane

Formula for Cholesterol
Figure 2.19

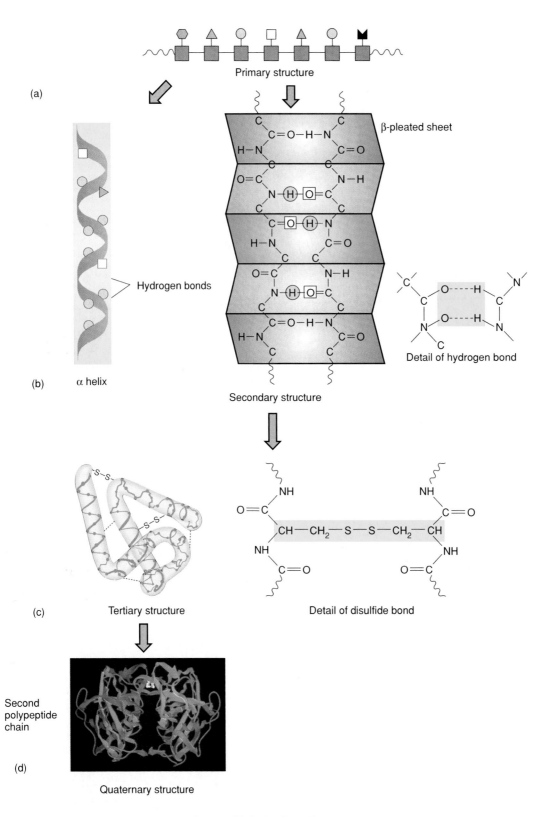

(a)

Primary structure

β-pleated sheet

(b) α helix

Hydrogen bonds

Detail of hydrogen bond

Secondary structure

(c) Tertiary structure

Detail of disulfide bond

Second polypeptide chain

(d)

Quaternary structure

Stages in the Formation of a Functioning Globular Protein
Figure 2.22

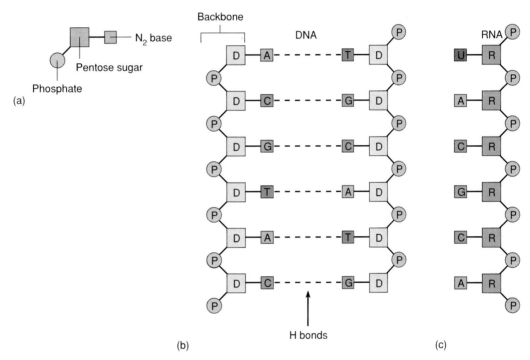

Basic Structure of Nucleic Acids
Figure 2.23

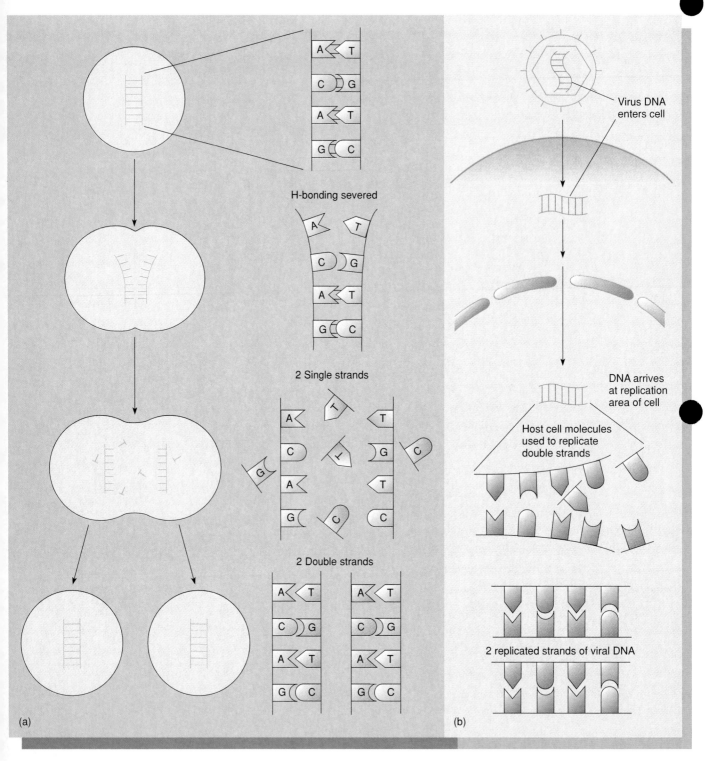

(a)

A ◄ T
C G
A ◄ T
G C

H-bonding severed

A T
C G
A ◄ T
G C

2 Single strands

A
C
A
G

T
T
G C
T
C

T
G
C

2 Double strands

A ◄ T
C G
A ◄ T
G C

A ◄ T
C G
A ◄ T
G C

(b)

Virus DNA enters cell

DNA arrives at replication area of cell

Host cell molecules used to replicate double strands

2 replicated strands of viral DNA

Replication of DNA
Figure 2.26

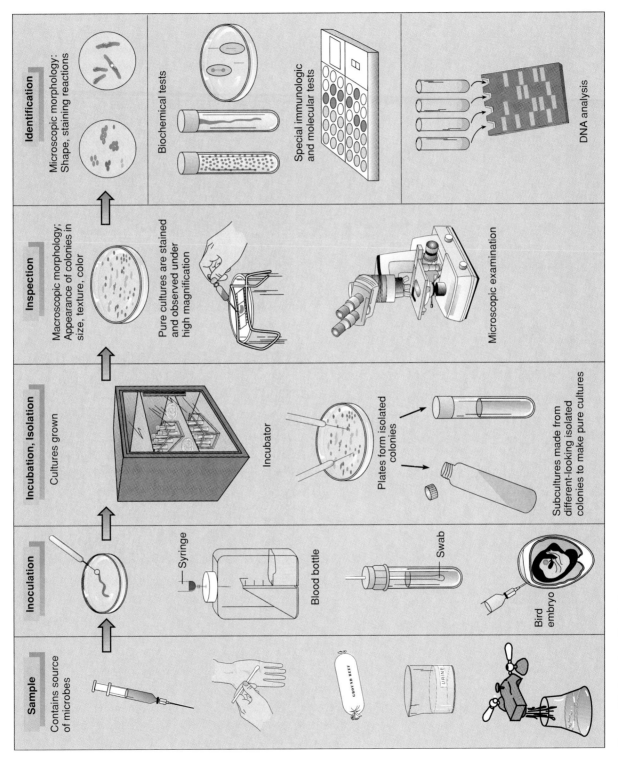

General Laboratory Techniques
Figure 3.1

Sample
Contains source of microbes

Inoculation
Syringe
Blood bottle
Swab
Bird embryo

Incubation, Isolation
Cultures grown
Incubator
Plates form isolated colonies
Subcultures made from different-looking isolated colonies to make pure cultures

Inspection
Macroscopic morphology; Appearance of colonies in size, texture, color
Pure cultures are stained and observed under high magnification
Microscopic examination

Identification
Microscopic morphology: Shape, staining reactions
Biochemical tests
Special immunologic and molecular tests
DNA analysis

13

Steps in a Streak Plate

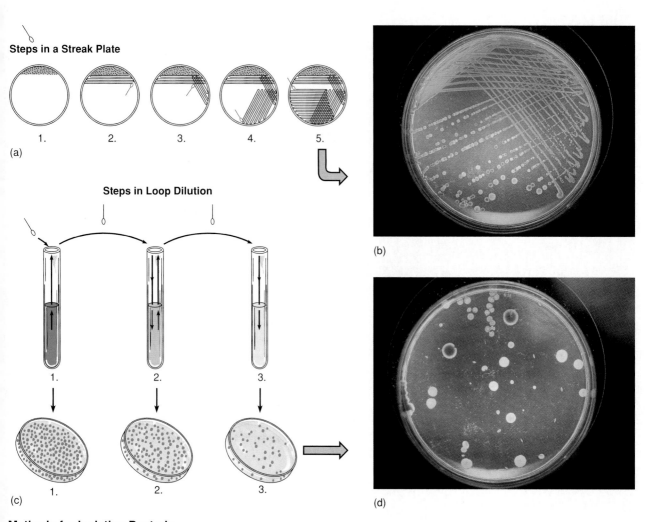

1.   2.   3.   4.   5.

(a)

Steps in Loop Dilution

1.   2.   3.

1.   2.   3.

(c)

(b)

(d)

Methods for Isolating Bacteria
Figure 3.4

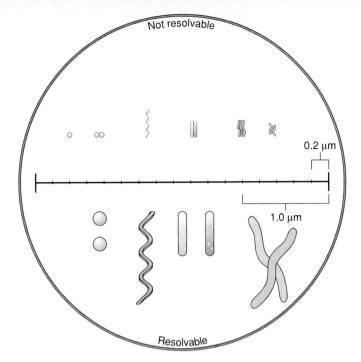

Resolvable Cells
Figure 3.19

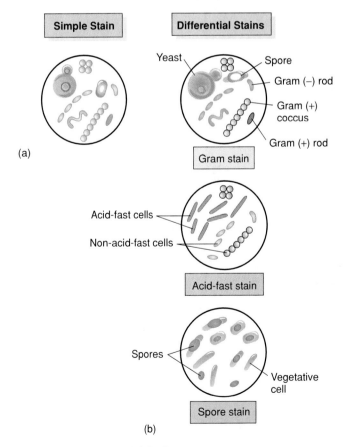

Simple and Differential Stains
Figure 3.27

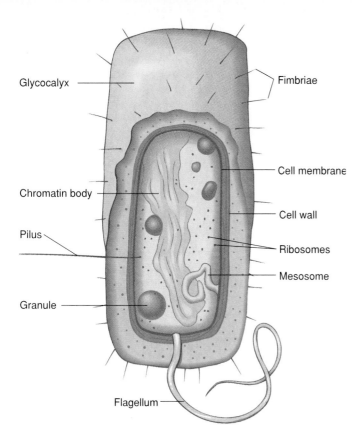

Glycocalyx

Fimbriae

Cell membrane

Chromatin body

Cell wall

Pilus

Ribosomes

Mesosome

Granule

Flagellum

(a)

Structural Features of Typical Rod-Shaped Bacterium
Figure 4.1a

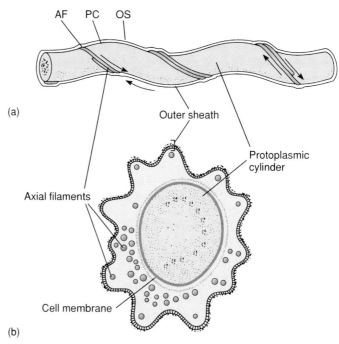

AF   PC   OS

(a)

Outer sheath

Axial filaments

Protoplasmic
cylinder

Cell membrane

(b)

Orientation of Axial Filaments on the Spirochete Cell
Figure 4.7

16

	Microscopic Appearance of Cell		Chemical Reaction in Cell Wall (very magnified view)	
Step	Gram (+)	Gram (−)	Gram (+)	Gram (−)
1. Crystal violet				Both cell walls affix the dye
2. Gram's iodine			Dye crystals trapped in wall	No effect of iodine
3. Alcohol			Crystals remain in cell wall	Cell wall partially dissolved, loses dye
4. Safranin (red dye)			Red dye has no effect	Red dye stains the colorless cell

Gram Stain Technique and Theory
Microfile Box Figure 4.1

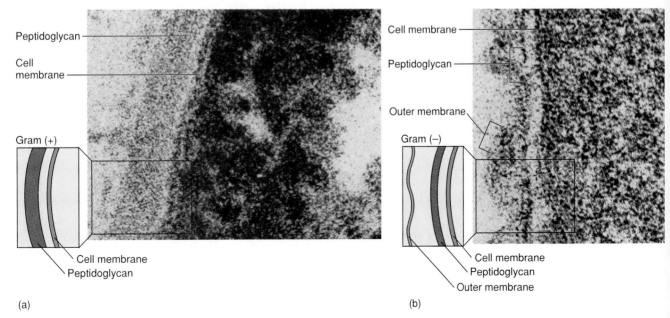

(a)

(b)

Peptidoglycan

Cell membrane

Gram (+)

Cell membrane
Peptidoglycan

Cell membrane

Peptidoglycan

Outer membrane

Gram (−)

Cell membrane
Peptidoglycan
Outer membrane

Comparison of Gram-Positive and Gram-Negative Envelopes
Figure 4.15

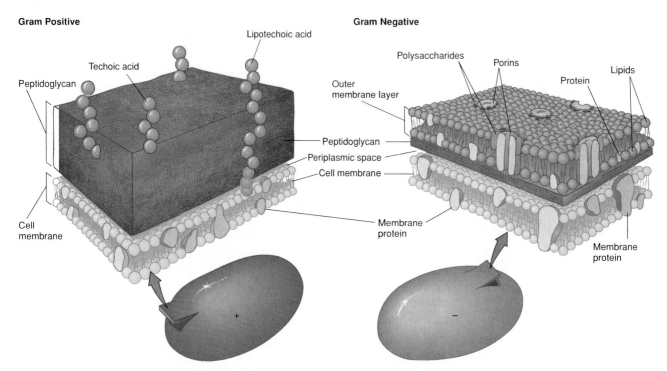

Gram Positive

Lipotechoic acid

Techoic acid

Peptidoglycan

Cell
membrane

Gram Negative

Polysaccharides

Porins

Protein

Lipids

Outer
membrane layer

Peptidoglycan

Periplasmic space

Cell membrane

Membrane
protein

Membrane
protein

Comparison of Gram-Positive and Gram-Negative Cell Walls
Figure 4.16

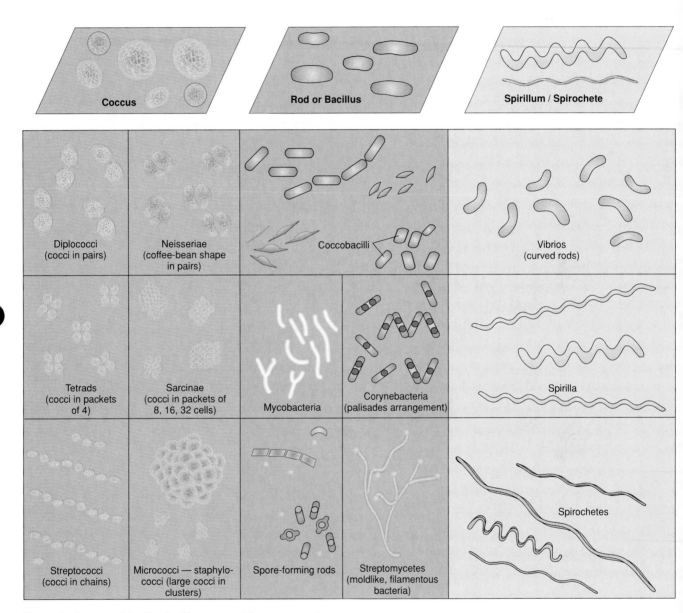

Bacteria Arranged by Basic Shapes and Arrangements
Figure 4.22

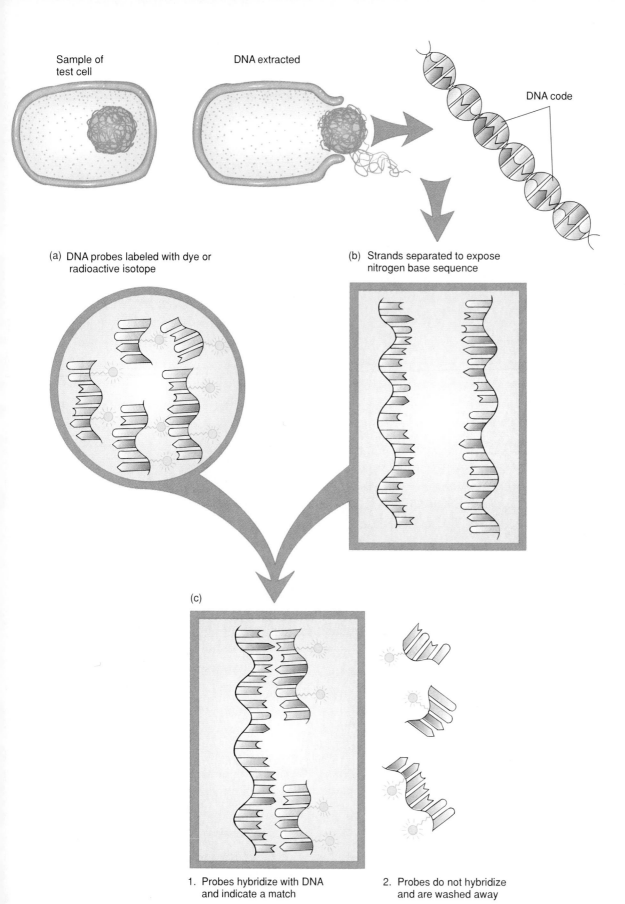

Sample of
test cell

DNA extracted

DNA code

(a) DNA probes labeled with dye or
radioactive isotope

(b) Strands separated to expose
nitrogen base sequence

(c)

1. Probes hybridize with DNA
and indicate a match

2. Probes do not hybridize
and are washed away

DNA Hybridization Using Probes
Figure 4.28

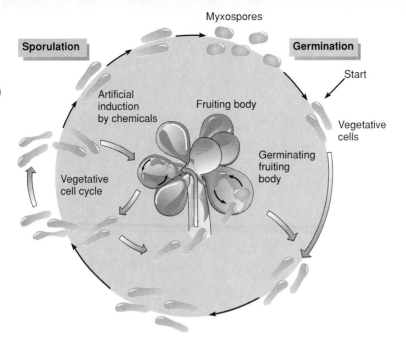

Sporulation

Myxospores

Germination

Start

Artificial induction by chemicals

Fruiting body

Vegetative cells

Germinating fruiting body

Vegetative cell cycle

The Life Cycle of Myxobacterium
Figure 4.36

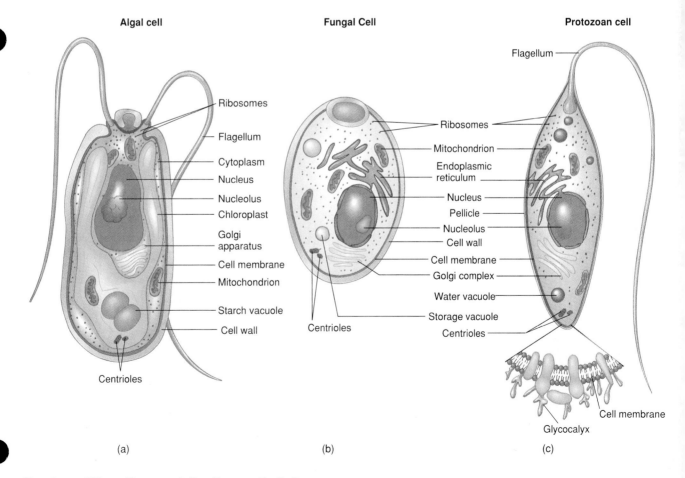

Algal cell

Fungal Cell

Protozoan cell

Flagellum

Ribosomes
Flagellum
Cytoplasm
Nucleus
Nucleolus
Chloroplast
Golgi apparatus
Cell membrane
Mitochondrion
Starch vacuole
Cell wall

Centrioles

Ribosomes
Mitochondrion
Endoplasmic reticulum
Nucleus
Pellicle
Nucleolus
Cell wall
Cell membrane
Golgi complex
Water vacuole
Storage vacuole
Centrioles

Centrioles

Cell membrane

Glycocalyx

(a)

(b)

(c)

Structure of Three Representative Eucaryotic Cells
Figure 5.2

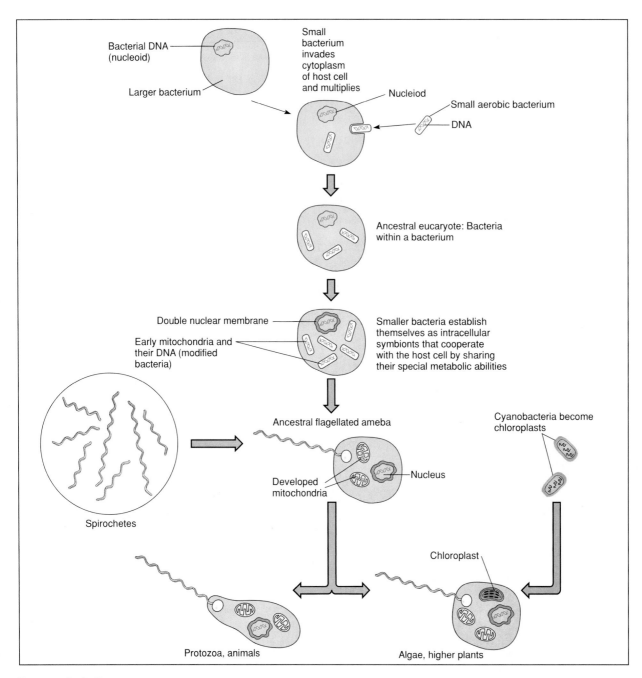

Eucaryotic Cells
Microfile Box Figure 5.1

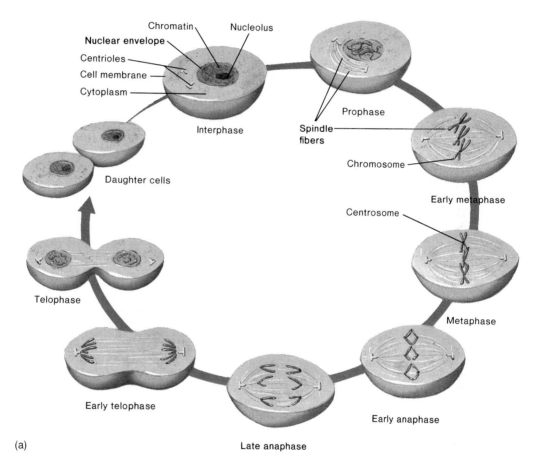

Changes in the Cell and Nucleus That Accompany Mitosis
Figure 5.6a

(a)

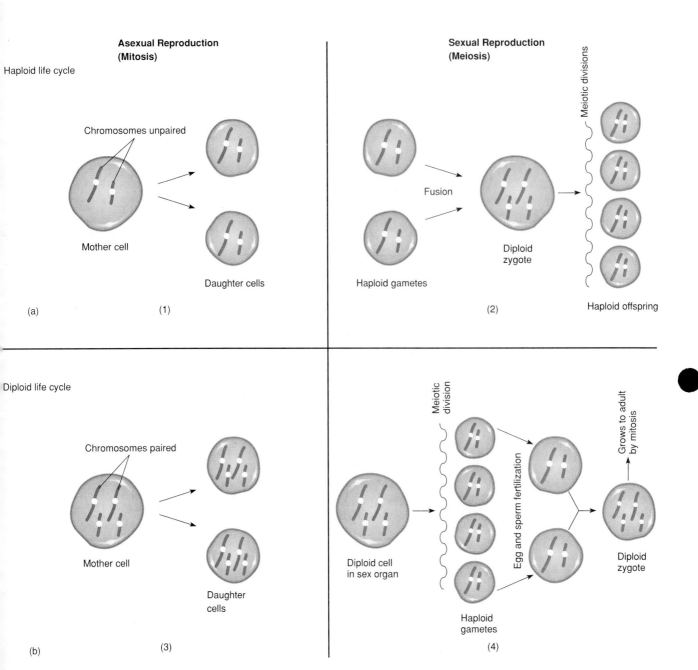

Asexual Reproduction (Mitosis)

Sexual Reproduction (Meiosis)

Haploid life cycle

Chromosomes unpaired

Mother cell

Daughter cells

(a)　(1)

Fusion

Haploid gametes

Diploid zygote

Meiotic divisions

Haploid offspring

(2)

Diploid life cycle

Chromosomes paired

Mother cell

Daughter cells

(b)　(3)

Meiotic division

Diploid cell in sex organ

Egg and sperm fertilization

Haploid gametes

Grows to adult by mitosis

Diploid zygote

(4)

Schematic of Haploid versus Diploid Life Cycles
Figure 5.7

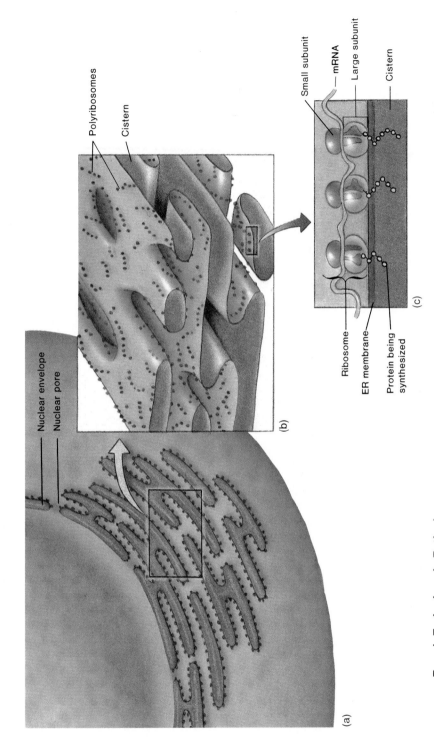

Rough Endoplasmic Reticulum
Figure 5.8

25

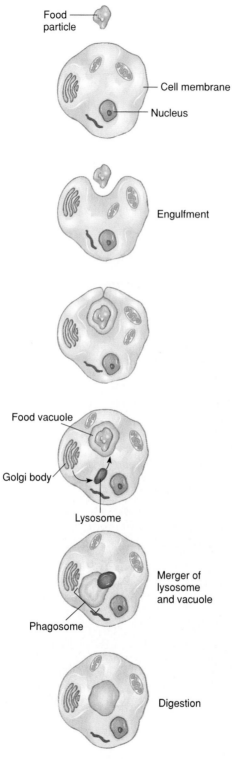

Food particle

Cell membrane

Nucleus

Engulfment

Food vacuole

Golgi body

Lysosome

Merger of lysosome and vacuole

Phagosome

Digestion

The Origin and Action of Lysosomes in Phagocytosis
Figure 5.11

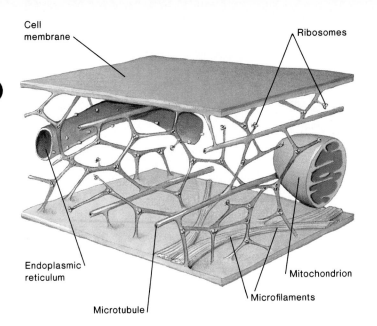

Cell membrane

Ribosomes

Endoplasmic reticulum

Mitochondrion

Microtubule

Microfilaments

Model of the Cytoskeleton
Figure 5.14

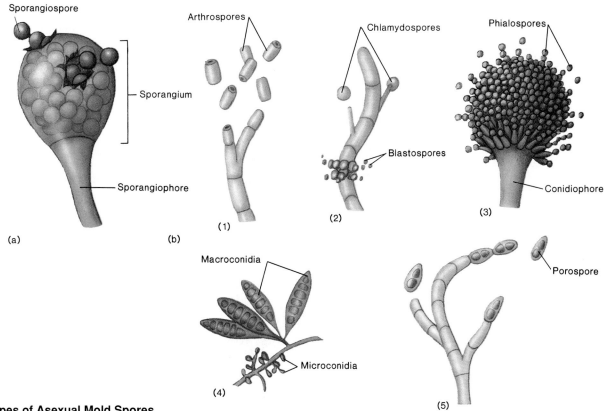

Sporangiospore

Arthrospores

Chlamydospores

Phialospores

Sporangium

Blastospores

Sporangiophore

Conidiophore

(a)

(b)

(1)

(2)

(3)

Macroconidia

Porospore

Microconidia

(4)

(5)

Types of Asexual Mold Spores
Figure 5.19

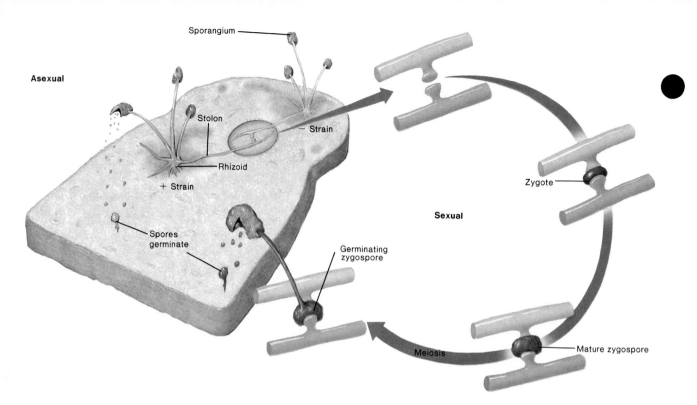

Formation of Zygospores in *Rhizopus stolonifer*
Figure 5.20

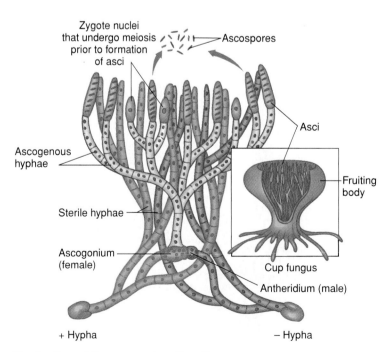

Production of Ascopores in a Cup Fungus
Figure 5.21

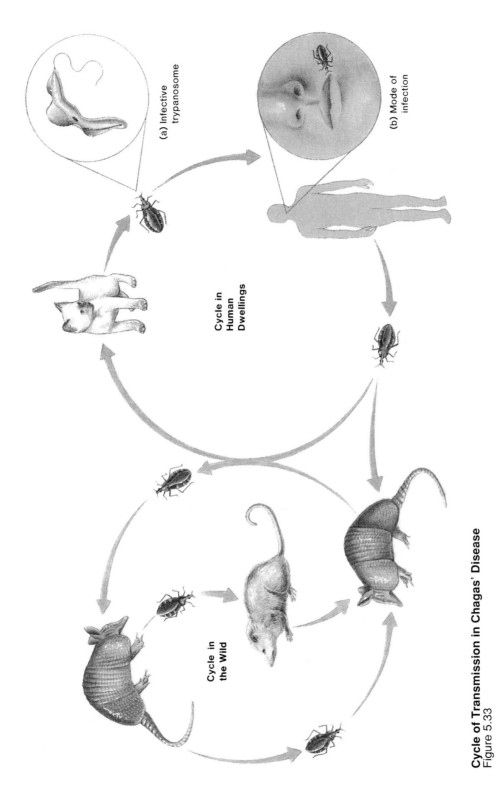

(a) Infective trypanosome

(b) Mode of infection

Cycle in Human Dwellings

Cycle in the Wild

Cycle of Transmission in Chagas' Disease
Figure 5.33

29

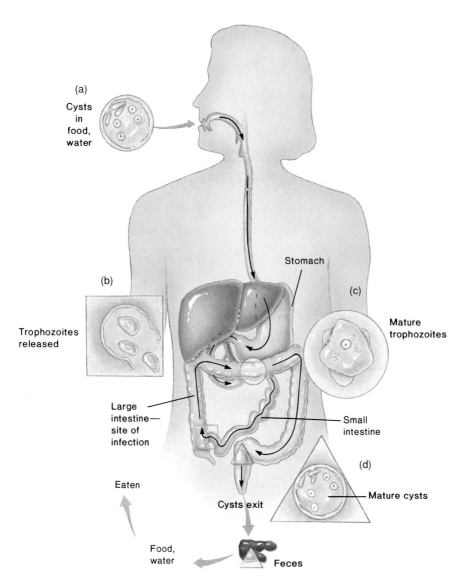

Stages in the Infection and Transmission of Amebic Dysentery
Figure 5.34

(a) Cysts in food, water

(b) Trophozoites released

Stomach

(c) Mature trophozoites

Large intestine—site of infection

Small intestine

(d) Mature cysts

Eaten

Cysts exit

Food, water

Feces

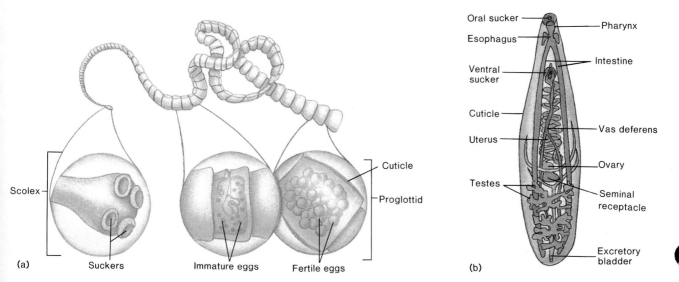

Parasitic Flatworms
Figure 5.35

(a) Scolex — Suckers — Immature eggs — Fertile eggs — Cuticle — Proglottid

(b) Oral sucker — Pharynx — Esophagus — Intestine — Ventral sucker — Cuticle — Vas deferens — Uterus — Ovary — Testes — Seminal receptacle — Excretory bladder

Naked nucleocapsid virus

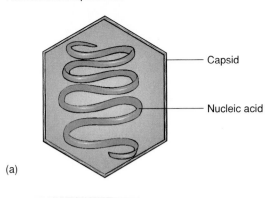

Capsid

Nucleic acid

(a)

Enveloped virus

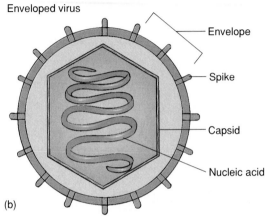

Envelope

Spike

Capsid

Nucleic acid

(b)

Generalized Structure of Viruses
Figure 6.3

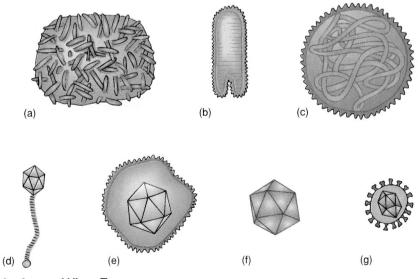

(a)　　　(b)　　　(c)

(d)　　(e)　　(f)　　(g)

An Array of Virus Types
Figure 6.9

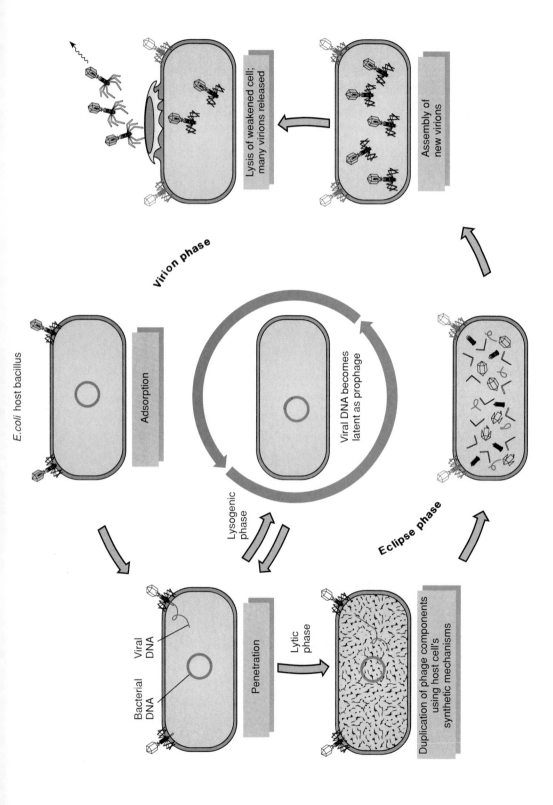

E.coli host bacillus

Adsorption

Bacterial DNA

Viral DNA

Penetration

Lytic phase

Duplication of phage components using host cell's synthetic mechanisms

Lysogenic phase

Viral DNA becomes latent as prophage

Eclipse phase

Virion phase

Assembly of new virions

Lysis of weakened cell; many virions released

Events in the Multiplication Cycle of T-Even Bacteriophages
Figure 6.10

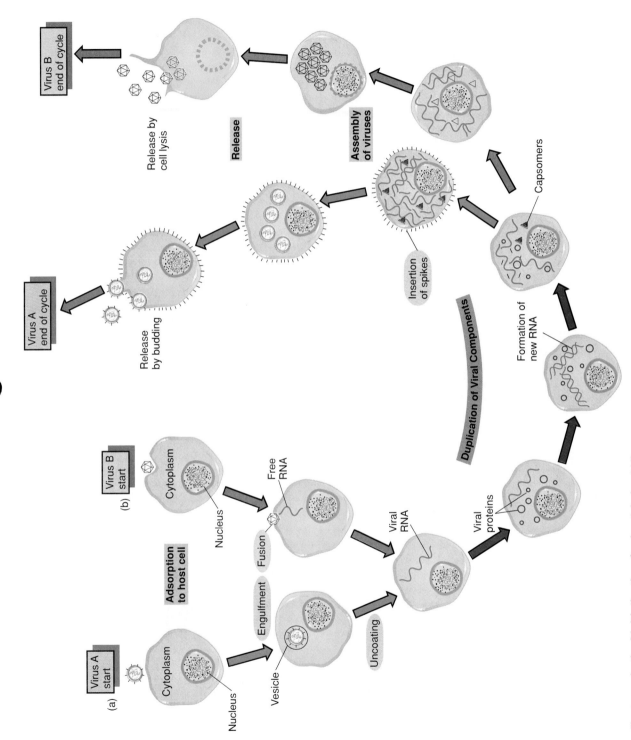

Features in the Multiplication Cycle of Animal Viruses
Figure 6.15

(a) Virus A start

Cytoplasm

Nucleus

Adsorption to host cell

Engulfment

Fusion

Vesicle

Uncoating

(b) Virus B start

Cytoplasm

Nucleus

Free RNA

Viral RNA

Viral proteins

Duplication of Viral Components

Formation of new RNA

Capsomers

Insertion of spikes

Assembly of viruses

Release

Release by cell lysis

Virus B end of cycle

Release by budding

Virus A end of cycle

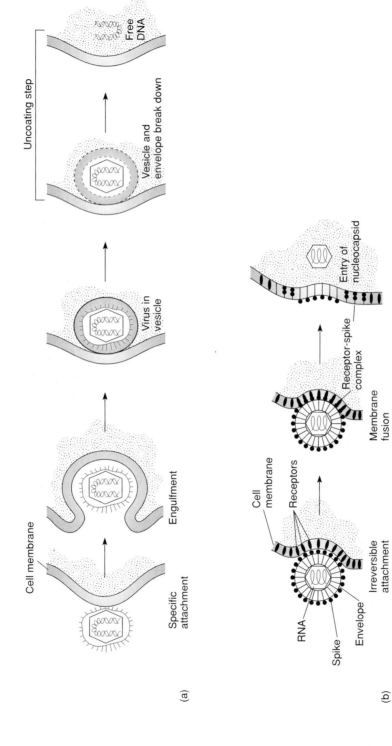

Uncoating step

Free DNA

Vesicle and envelope break down

Virus in vesicle

Cell membrane

Engulfment

Specific attachment

(a)

Cell membrane

Receptors

Entry of nucleocapsid

Receptor-spike complex

Membrane fusion

Irreversible attachment

RNA

Spike

Envelope

(b)

Two Principle Means by Which Animal Viruses Penetrate
Figure 6.17

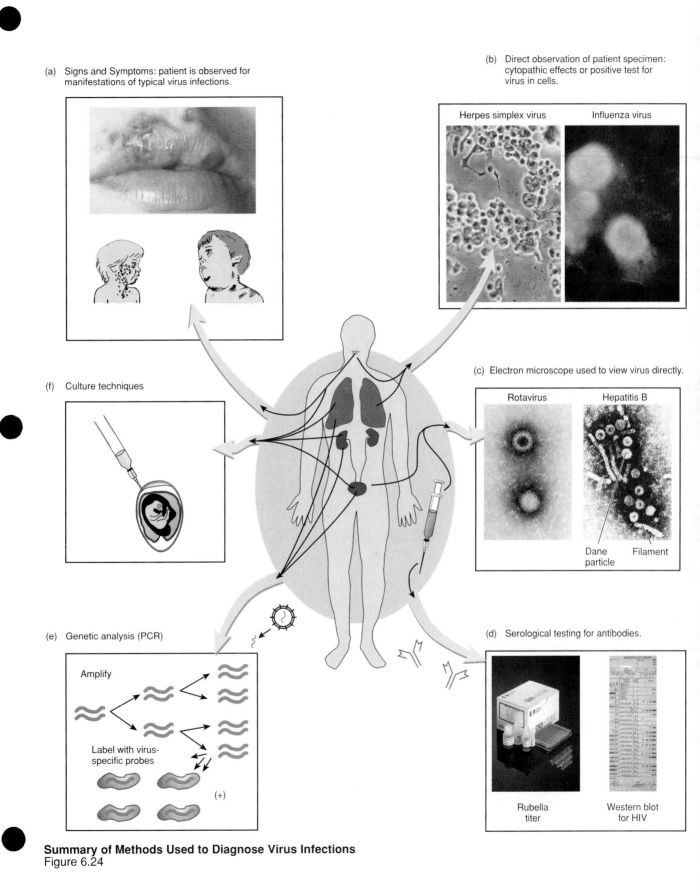

(a) Signs and Symptoms: patient is observed for manifestations of typical virus infections.

(b) Direct observation of patient specimen: cytopathic effects or positive test for virus in cells.

Herpes simplex virus   Influenza virus

(c) Electron microscope used to view virus directly.

Rotavirus   Hepatitis B

Dane particle   Filament

(f) Culture techniques

(e) Genetic analysis (PCR)

Amplify

Label with virus-specific probes

(+)

(d) Serological testing for antibodies.

Rubella titer   Western blot for HIV

Summary of Methods Used to Diagnose Virus Infections
Figure 6.24

Digestion in Bacteria and Fungi

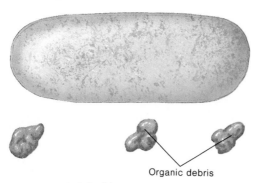

(a) Walled cell is inflexible.

Organic debris

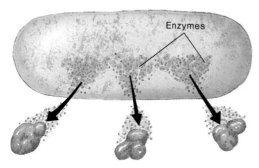

Enzymes

(b) Enzymes are transported across the wall.

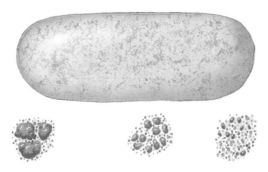

(c) Enzymes hydrolyze the bonds on nutrients.

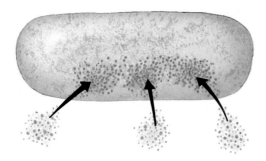

(d) Smaller molecules are transported into the cytoplasm.

Extracellular Digestion in a Saprobe with a Cell Wall
Figure 7.3

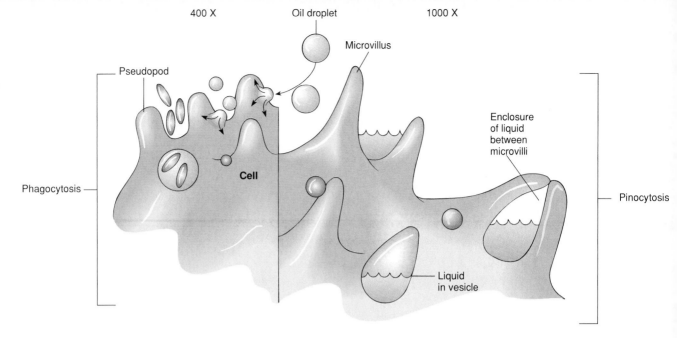

400 X

Oil droplet

Microvillus

1000 X

Pseudopod

Enclosure
of liquid
between
microvilli

Cell

Phagocytosis

Pinocytosis

Liquid
in vesicle

Endocytosis
Figure 7.9

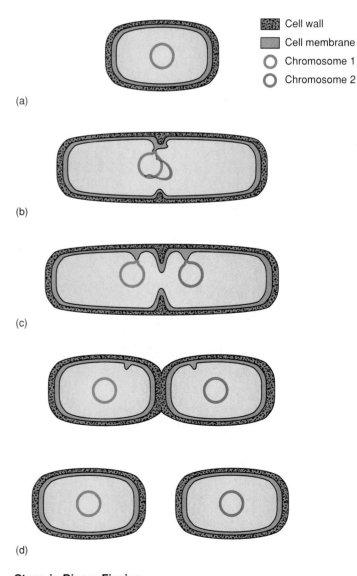

Cell wall

Cell membrane

Chromosome 1

Chromosome 2

(a)

(b)

(c)

(d)

Steps in Binary Fission
Figure 7.15

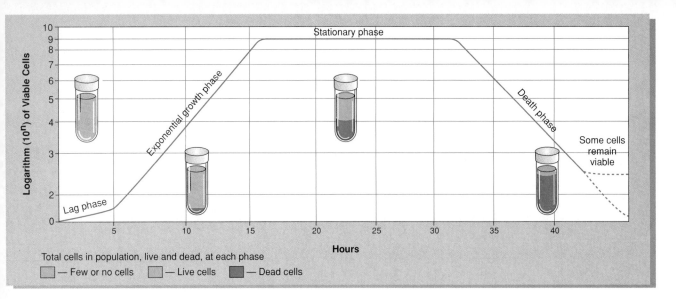

The Growth Curve
Figure 7.17

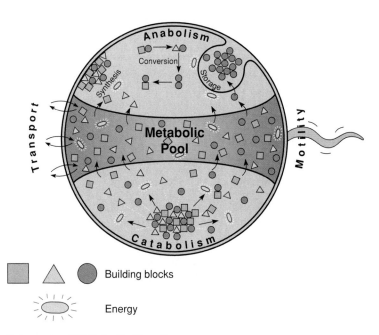

Summary of Metabolic Functions
Figure 8.1

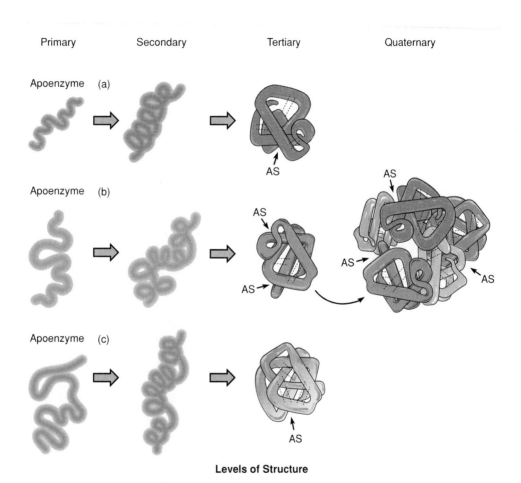

Primary   Secondary   Tertiary   Quaternary

Apoenzyme   (a)

AS

Apoenzyme   (b)

AS

AS

AS

AS

AS

AS

Apoenzyme   (c)

AS

Levels of Structure

How the Active Site and Specificity of the Apoenzyme Arise
Figure 8.4

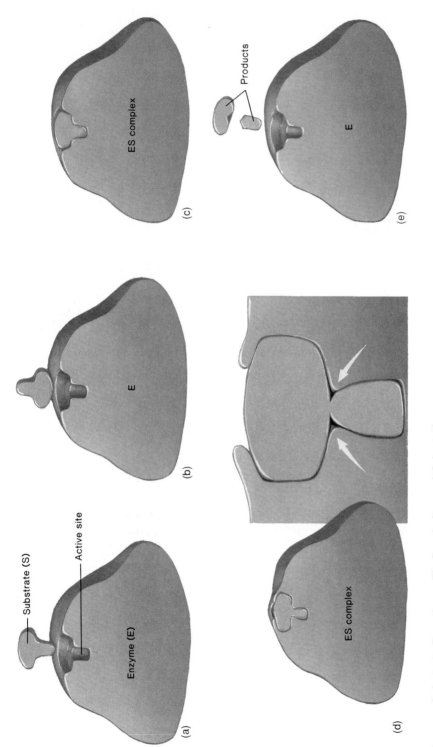

Enzyme-Substrate Reactions: Fit, Proximity, and Orientation
Figure 8.5

40

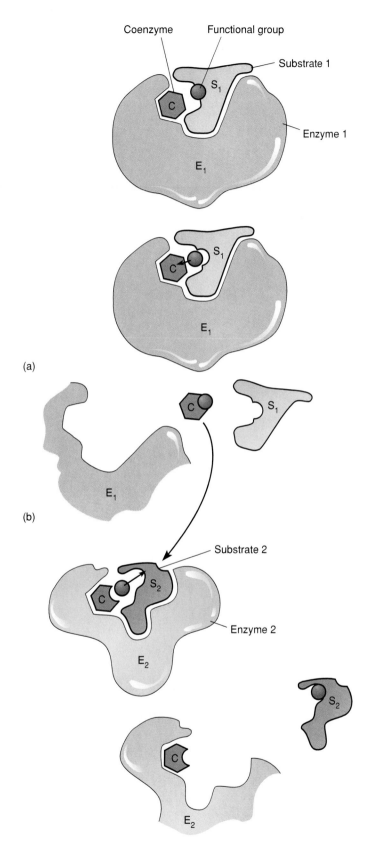

Coenzyme   Functional group

Substrate 1

S_1

Enzyme 1

E_1

S_1

C

E_1

(a)

C

S_1

E_1

(b)

Substrate 2

S_2

C

Enzyme 2

E_2

S_2

C

E_2

The Carrier Functions of Coenzymes
Figure 8.6

41

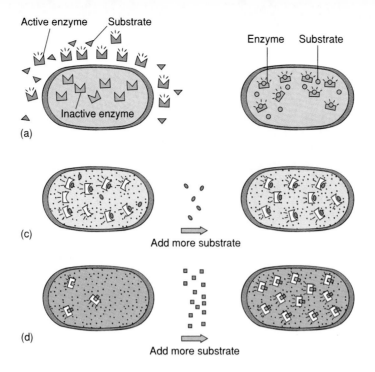

Types of Enzymes
Figure 8.7

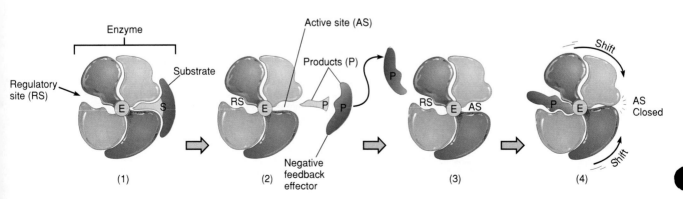

Enzyme Control by Negative Feedback in System
Figure 8.10

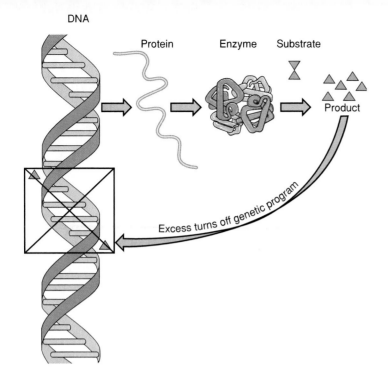

Feedback/Enzyme Repression
Figure 8.11

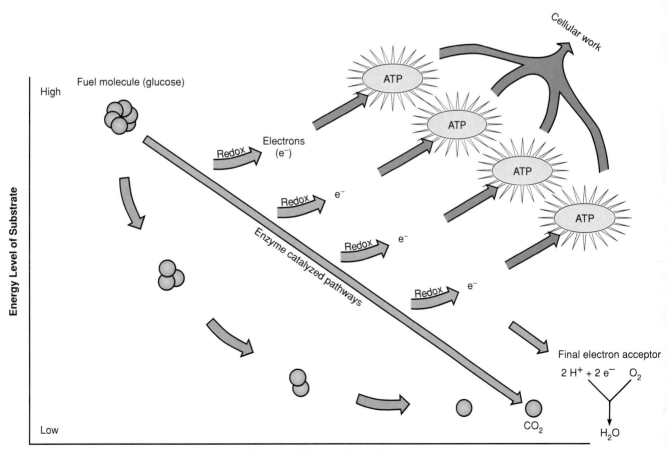

A Simplified View of the Cell's Energy Machine
Figure 8.12

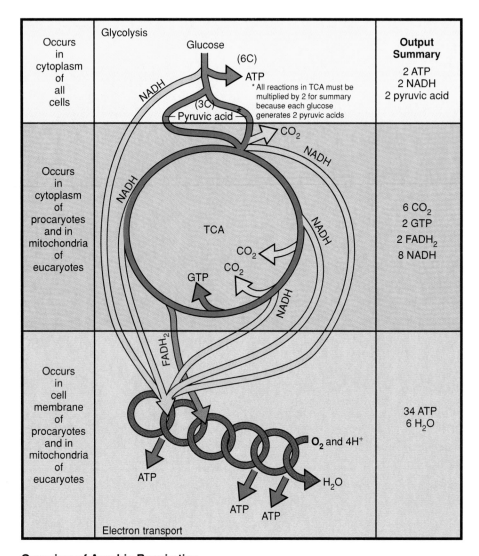

Occurs in cytoplasm of all cells	Glycolysis Glucose (6C) ATP NADH (3C) Pyruvic acid * All reactions in TCA must be multiplied by 2 for summary because each glucose generates 2 pyruvic acids	Output Summary 2 ATP 2 NADH 2 pyruvic acid
Occurs in cytoplasm of procaryotes and in mitochondria of eucaryotes	CO_2 NADH NADH NADH TCA CO_2 CO_2 GTP NADH	6 CO_2 2 GTP 2 $FADH_2$ 8 NADH
Occurs in cell membrane of procaryotes and in mitochondria of eucaryotes	$FADH_2$ ATP O_2 and 4H$^+$ H_2O ATP ATP Electron transport	34 ATP 6 H_2O

Overview of Aerobic Respiration
Figure 8.18

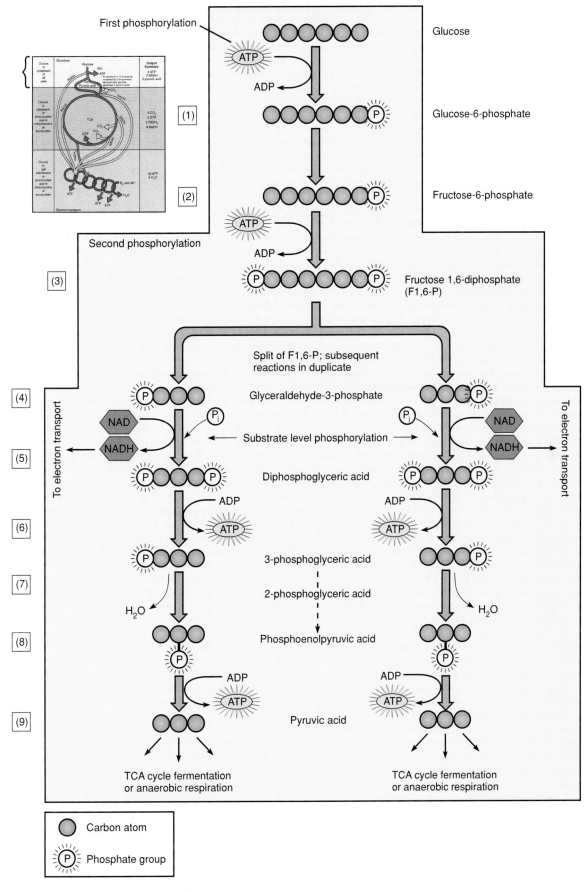

First phosphorylation

Glucose

ATP

ADP

(1)

Glucose-6-phosphate

Fructose-6-phosphate

(2)

Second phosphorylation

ATP

ADP

(3)

Fructose 1,6-diphosphate
(F1,6-P)

Split of F1,6-P; subsequent
reactions in duplicate

(4)

Glyceraldehyde-3-phosphate

To electron transport

NAD

P_i

Substrate level phosphorylation

P_i

NAD

To electron transport

(5)

NADH

NADH

Diphosphoglyceric acid

ADP

ATP

(6)

ADP

ATP

3-phosphoglyceric acid

(7)

2-phosphoglyceric acid

H_2O

H_2O

(8)

Phosphoenolpyruvic acid

ADP

ATP

(9)

ADP

ATP

Pyruvic acid

TCA cycle fermentation
or anaerobic respiration

TCA cycle fermentation
or anaerobic respiration

Carbon atom

P Phosphate group

The Reactions of the Glycolysis System
Figure 8.19

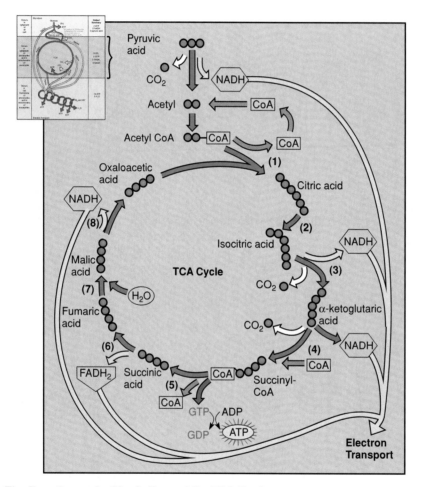

The Reactions of a Single Turn of the TCA Cycle
Figure 8.21

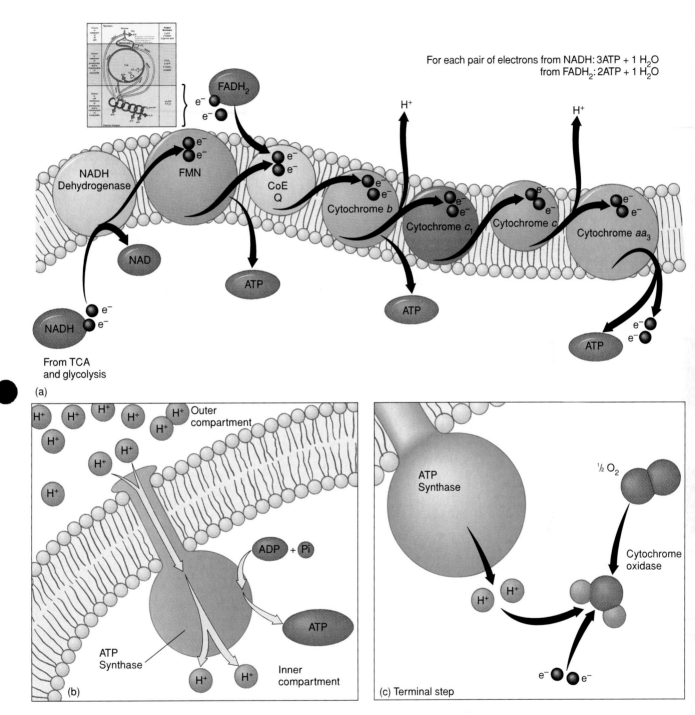

For each pair of electrons from NADH: 3ATP + 1 H$_2$O
from FADH$_2$: 2ATP + 1 H$_2$O

NADH Dehydrogenase
FMN
CoE Q
Cytochrome b
Cytochrome c_1
Cytochrome c
Cytochrome aa_3

e$^-$
FADH$_2$

H$^+$

NAD
ATP
ATP

NADH
From TCA and glycolysis
(a)

H$^+$ Outer compartment
ADP + Pi
ATP
ATP Synthase
H$^+$
Inner compartment
(b)

ATP Synthase
½ O$_2$
Cytochrome oxidase
H$^+$
e$^-$
(c) Terminal step

Electron Transport Chain
Figure 8.23

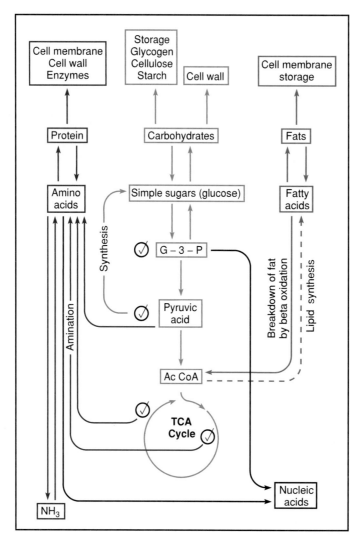

A Summary of Metabolic Interactions
Figure 8.26

Amination

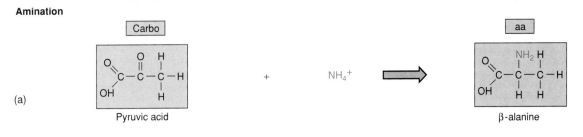

(a) Pyruvic acid + NH₄⁺ → β-alanine

Transamination

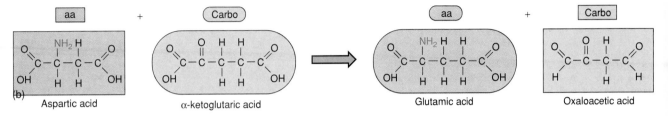

(b) Aspartic acid + α-ketoglutaric acid → Glutamic acid + Oxaloacetic acid

Deamination

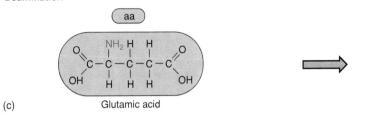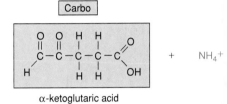

(c) Glutamic acid → α-ketoglutaric acid + NH₄⁺

Reactions That Produce and Convert Amino Acids
Figure 8.27

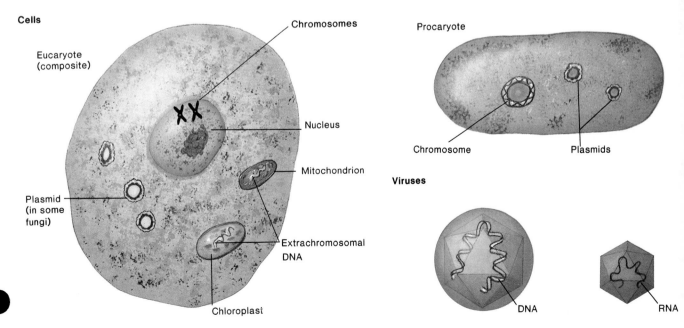

Location and Forms of the Genome in Microbes
Figure 9.2

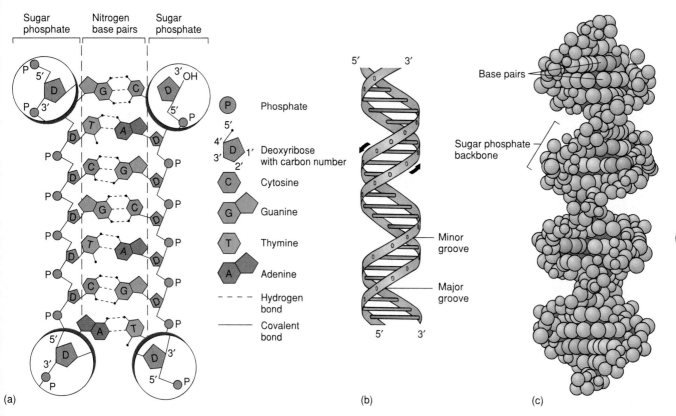

Three Views of DNA Structure
Figure 9.4

In figure (a), labels read:
Sugar phosphate | Nitrogen base pairs | Sugar phosphate

Legend:
- P Phosphate
- D (Deoxyribose with carbon number) 5′ 4′ 3′ 2′ 1′
- C Cytosine
- G Guanine
- T Thymine
- A Adenine
- - - - Hydrogen bond
- ——— Covalent bond

(a)

In figure (b), labels read: 5′ 3′ (top), Minor groove, Major groove, 5′ 3′ (bottom)

(b)

In figure (c), labels read: Base pairs, Sugar phosphate backbone

(c)

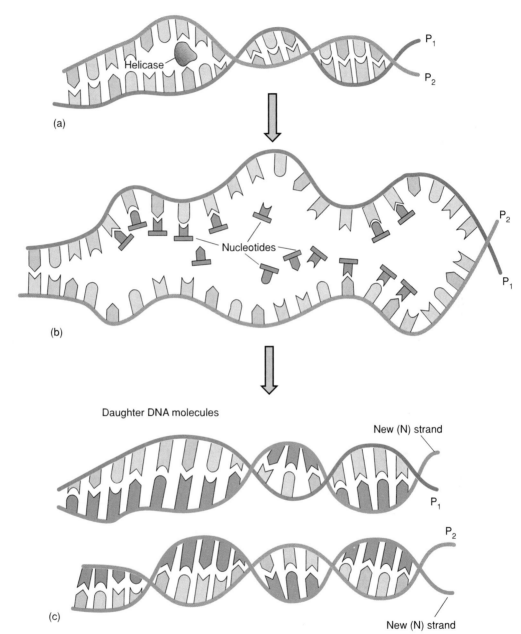

(a)

Helicase

P₁

P₂

(b)

Nucleotides

P₂

P₁

Daughter DNA molecules

New (N) strand

P₁

P₂

New (N) strand

(c)

Simplified Steps in Semiconservative Replication of DNA
Figure 9.6

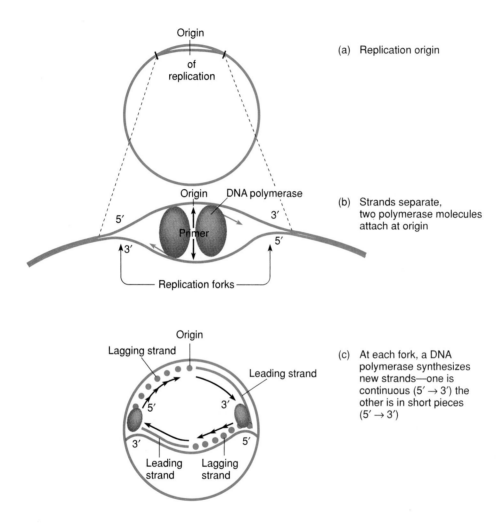

(a) Replication origin

(b) Strands separate, two polymerase molecules attach at origin

(c) At each fork, a DNA polymerase synthesizes new strands—one is continuous (5′ → 3′) the other is in short pieces (5′ → 3′)

The Bacterial Replication: a Model for DNA Synthesis
Figure 9.7

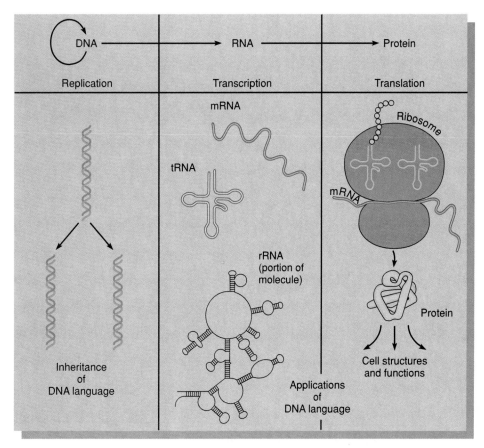

Flow of Genetic Information in Cells
Figure 9.10

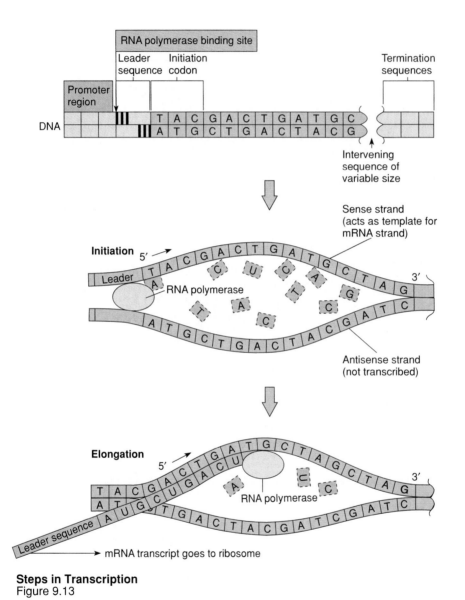

Steps in Transcription
Figure 9.13

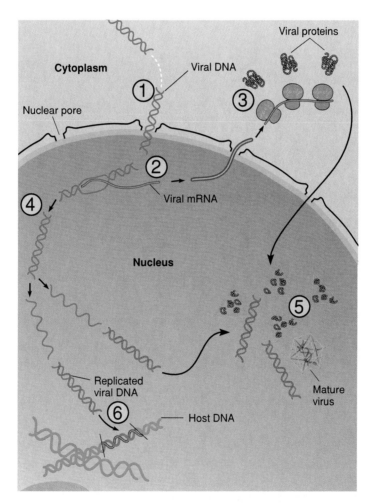

Genetic Stages in the Multiplication of DNA Viruses
Figure 9.19

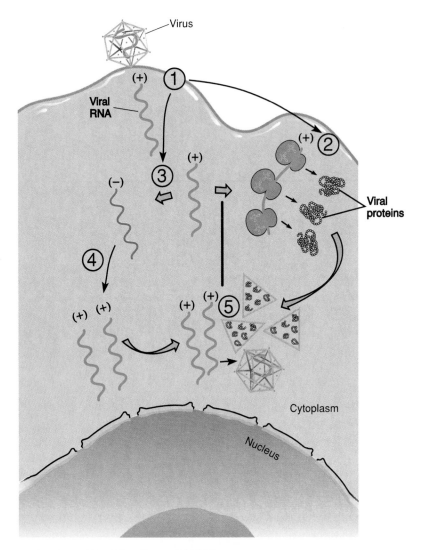

Replication of Positive-Sense RNA Viruses
Figure 9.20

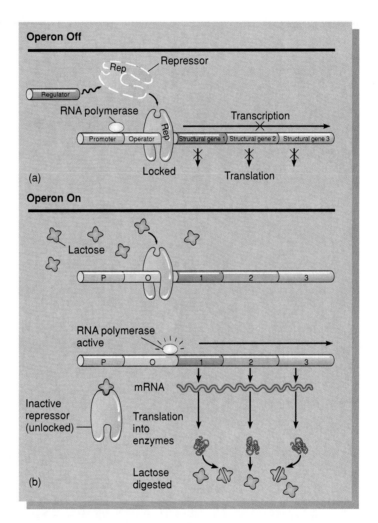

The Lactose Operon in Bacteria
Figure 9.21

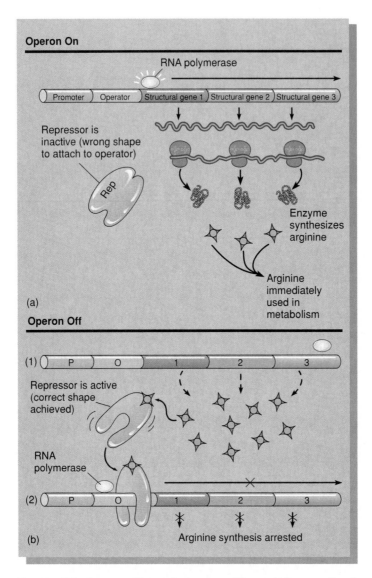

Repressible Operon: Control of a Gene Through Excess Nutrient
Figure 9.22

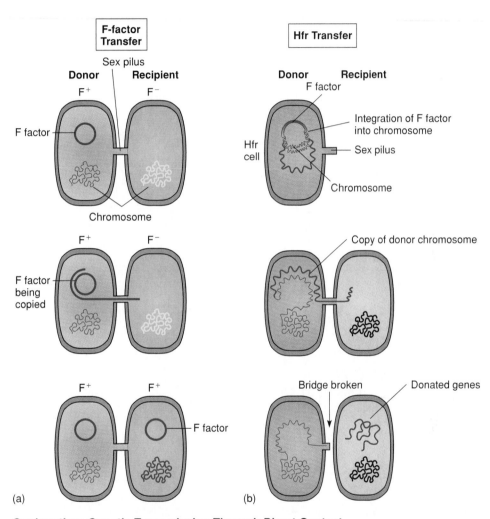

Conjugation: Genetic Transmission Through Direct Contact
Figure 9.27

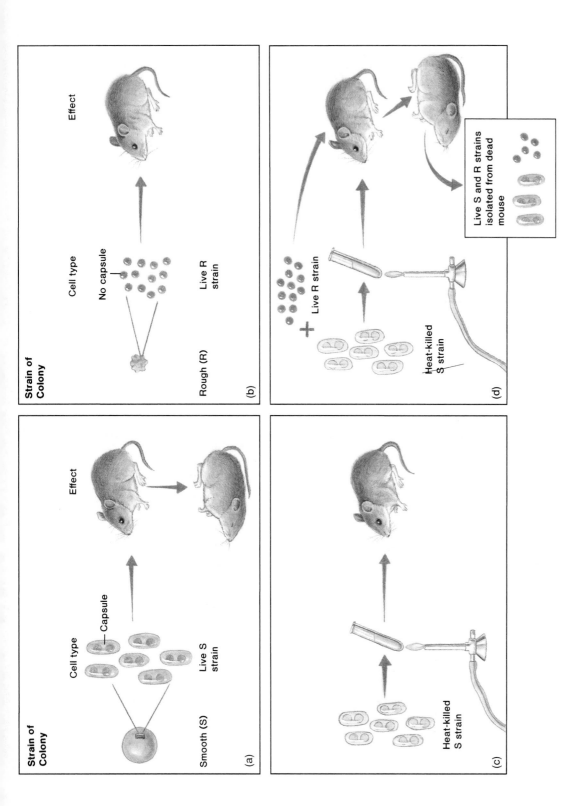

Strain of Colony

(a) Smooth (S)

Cell type — Capsule

Live S strain

Effect

(b) Rough (R)

Cell type — No capsule

Live R strain

Effect

(c) Heat-killed S strain

(d) Heat-killed S strain + Live R strain

Live S and R strains isolated from dead mouse

Griffith's Classic Experiment in Transformation
Figure 9.28

60

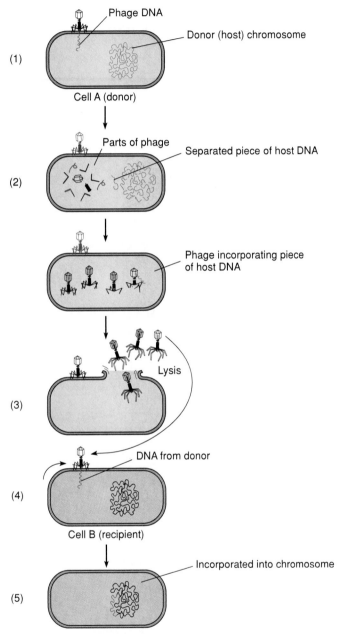

(1)

Phage DNA

Donor (host) chromosome

Cell A (donor)

(2)

Parts of phage

Separated piece of host DNA

Phage incorporating piece
of host DNA

(3)

Lysis

(4)

DNA from donor

Cell B (recipient)

(5)

Incorporated into chromosome

Cell survives and utilizes transduced DNA

Generalized Transduction: Genetic Transfer by Means of a Virus Carrier
Figure 9.29

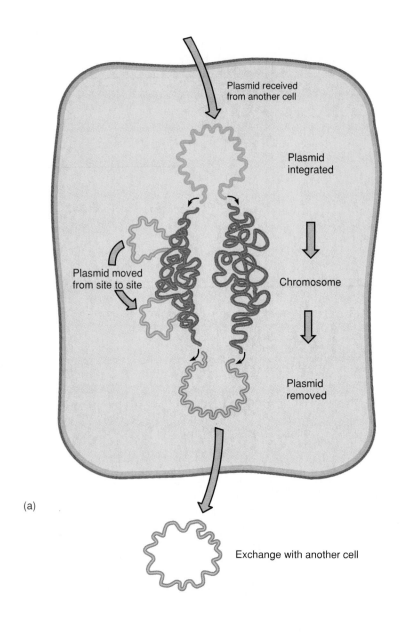

(a)

Plasmid received
from another cell

Plasmid
integrated

Plasmid moved
from site to site

Chromosome

Plasmid
removed

Exchange with another cell

Transposon

Palindrome

| | G A C G T | C T A C T G A | A C G T C | |
| | C T G C A | G A T G A C T | T G C A G | |

Palindrome

Palindrome

(b)

Transposons: Shifting Segments of the Genome
Figure 9.30

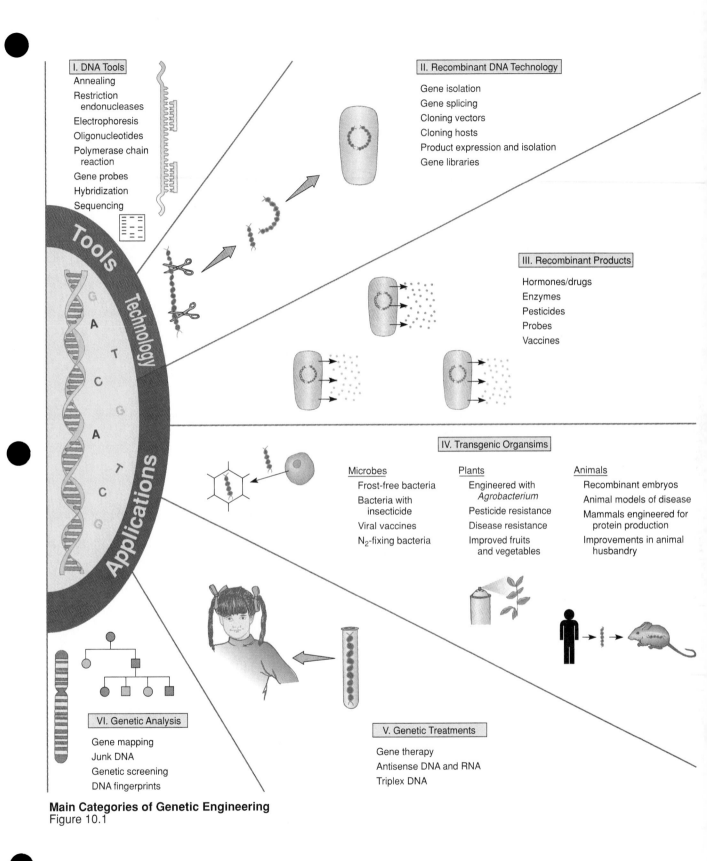

I. DNA Tools

Annealing
Restriction endonucleases
Electrophoresis
Oligonucleotides
Polymerase chain reaction
Gene probes
Hybridization
Sequencing

II. Recombinant DNA Technology

Gene isolation
Gene splicing
Cloning vectors
Cloning hosts
Product expression and isolation
Gene libraries

III. Recombinant Products

Hormones/drugs
Enzymes
Pesticides
Probes
Vaccines

IV. Transgenic Organsims

Microbes
Frost-free bacteria
Bacteria with insecticide
Viral vaccines
N_2-fixing bacteria

Plants
Engineered with *Agrobacterium*
Pesticide resistance
Disease resistance
Improved fruits and vegetables

Animals
Recombinant embryos
Animal models of disease
Mammals engineered for protein production
Improvements in animal husbandry

VI. Genetic Analysis

Gene mapping
Junk DNA
Genetic screening
DNA fingerprints

V. Genetic Treatments

Gene therapy
Antisense DNA and RNA
Triplex DNA

Tools
Technology
Applications

Main Categories of Genetic Engineering
Figure 10.1

63

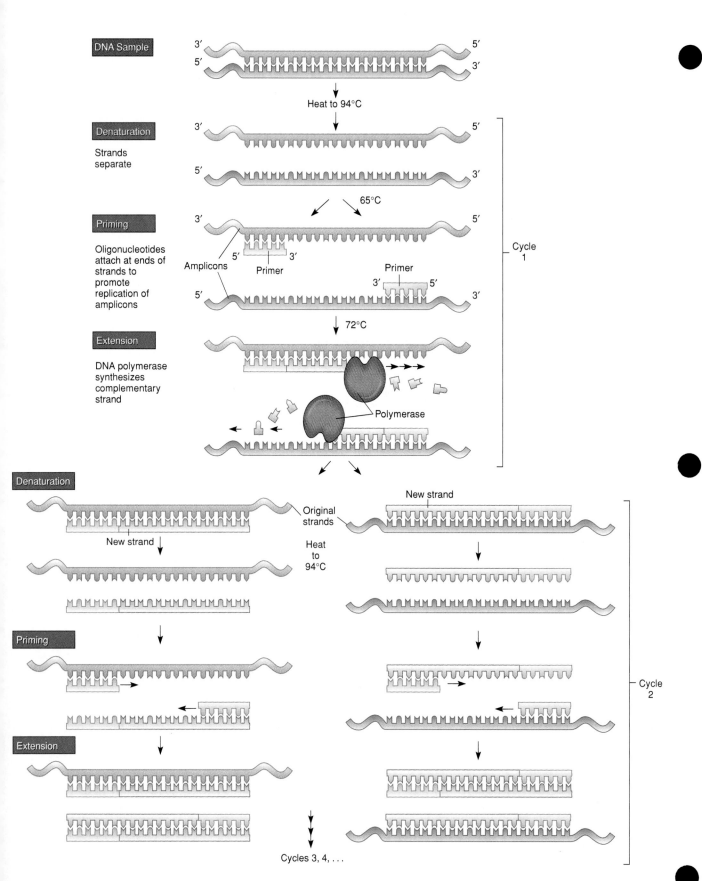

DNA Sample

3′ 5′
5′ 3′

Heat to 94°C

Denaturation

Strands
separate

3′ 5′

5′ 3′

65°C

Priming

Oligonucleotides
attach at ends of
strands to
promote
replication of
amplicons

3′ 5′

Amplicons
5′ Primer 3′

Primer
3′ 5′

5′ 3′

72°C

Extension

DNA polymerase
synthesizes
complementary
strand

Polymerase

Denaturation

New strand
Original
strands

New strand

Heat
to
94°C

Priming

Extension

Cycle
1

Cycle
2

Cycles 3, 4, . . .

Schematic of the Polymerase Chain Reaction
Figure 10.6

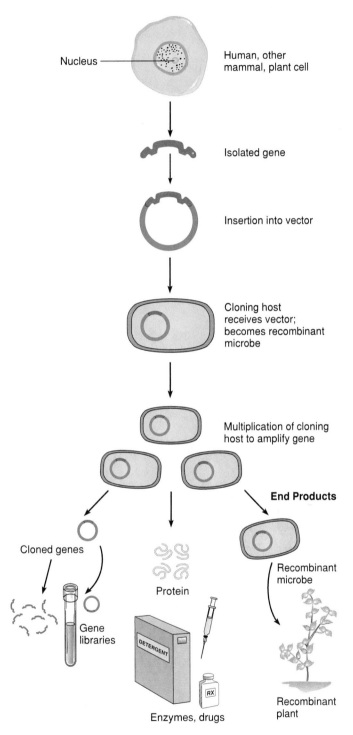

Nucleus —— Human, other mammal, plant cell

Isolated gene

Insertion into vector

Cloning host receives vector; becomes recombinant microbe

Multiplication of cloning host to amplify gene

End Products

Cloned genes

Gene libraries

Protein

DETERGENT

RX

Enzymes, drugs

Recombinant microbe

Recombinant plant

Strategy for the Applications of Gene Cloning in Genetic Engineering
Figure 10.7

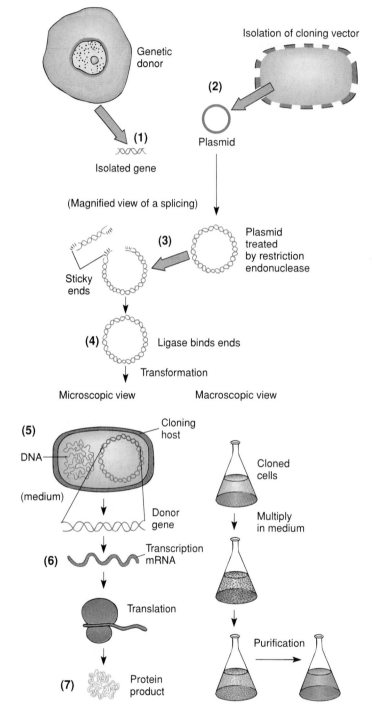

The General Steps in Recombinant DNA, Gene Cloning, and Product Retrieval
Figure 10.9

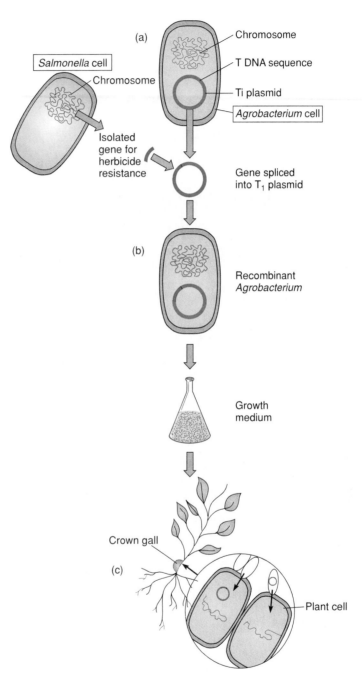

(a)

Salmonella cell

Chromosome

Isolated gene for herbicide resistance

Chromosome

T DNA sequence

Ti plasmid

Agrobacterium cell

Gene spliced into T₁ plasmid

(b)

Recombinant *Agrobacterium*

Growth medium

Crown gall

(c)

Plant cell

Bioengineering of Plants
Figure 10.12

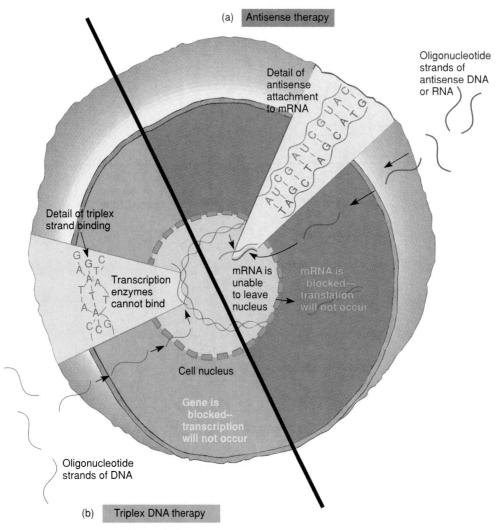

(a) Antisense therapy

Detail of antisense attachment to mRNA

Oligonucleotide strands of antisense DNA or RNA

A U C G A U C G U A C
T A G C T A G C A T G

Detail of triplex strand binding

Transcription enzymes cannot bind

mRNA is unable to leave nucleus

mRNA is blocked— translation will not occur

Cell nucleus

Gene is blocked— transcription will not occur

Oligonucleotide strands of DNA

(b) Triplex DNA therapy

Mechanisms of Antisense DNA and Triplex DNA
Figure 10.17

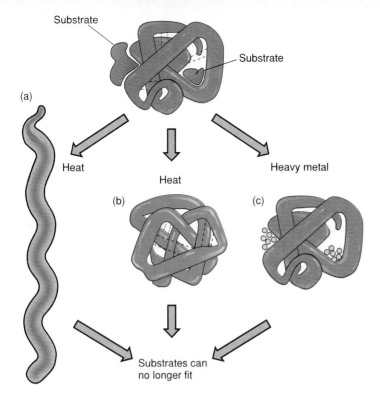

Modes of Action Affecting Protein Function
Figure 11.4

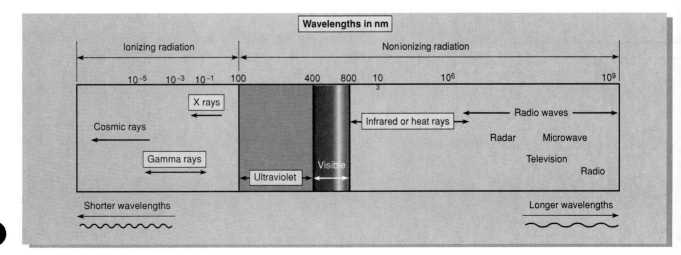

Electromagnetic Radiation Used in Chemical Control
Figure 11.7

Ionizing Radiation

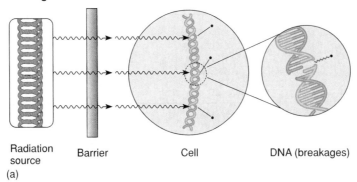

Radiation source
Barrier
Cell
DNA (breakages)

(a)

Nonionizing Radiation

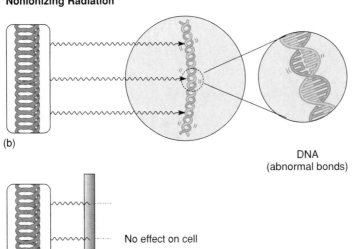

(b)

DNA
(abnormal bonds)

No effect on cell

Radiation source
Barrier

(c)

Cellular Effects of Irradiation
Figure 11.8

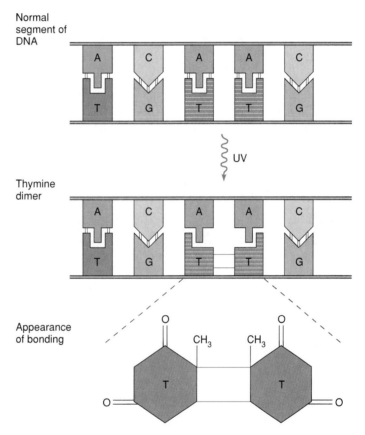

Normal segment of DNA

Thymine dimer

UV

Appearance of bonding

Formation of Pyrimidine Dimers by the Action of UV Radiation
Figure 11.11

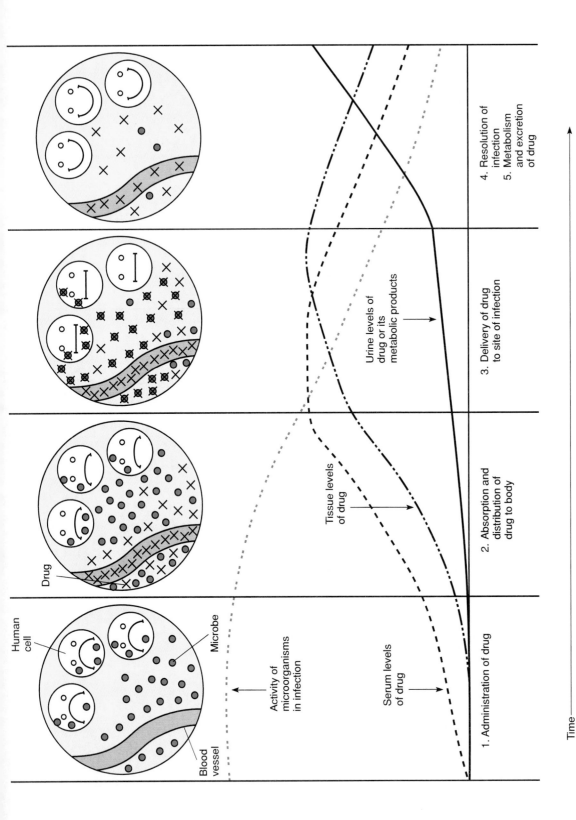

Human cell

Drug

Microbe

Blood vessel

Activity of microorganisms in infection

Serum levels of drug

Tissue levels of drug

Urine levels of drug or its metabolic products

1. Administration of drug

2. Absorption and distribution of drug to body

3. Delivery of drug to site of infection

4. Resolution of infection
5. Metabolism and excretion of drug

Time

The Course of Events in Chemotherapy
Figure 12.1

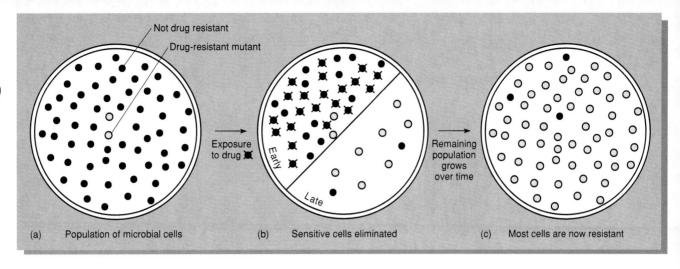

(a) Population of microbial cells
(b) Sensitive cells eliminated
(c) Most cells are now resistant

Not drug resistant
Drug-resistant mutant
Exposure to drug
Early
Late
Remaining population grows over time

Natural Selection and Drug Resistance
Figure 12.11

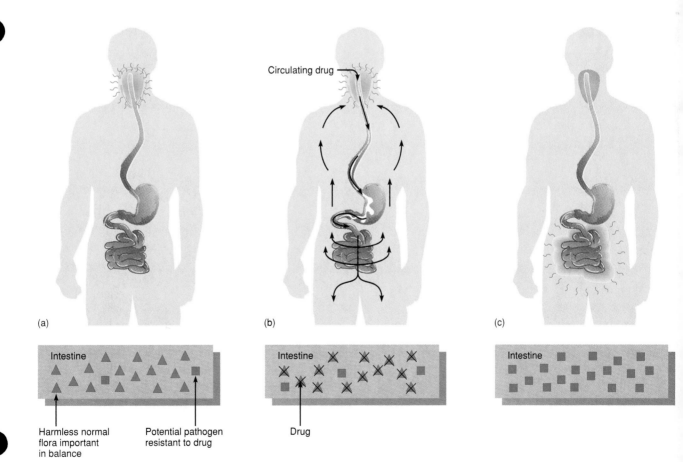

(a)
(b)
(c)

Circulating drug

| Intestine | Intestine | Intestine |

Harmless normal flora important in balance
Potential pathogen resistant to drug
Drug

The Role of Antimicrobics in Disrupting Flora and Causing Superinfections
Figure 12.20

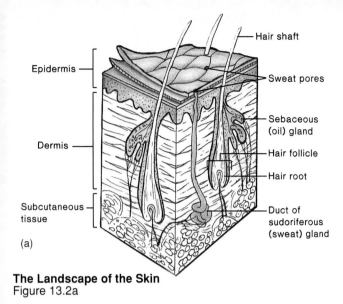

Epidermis

Dermis

Subcutaneous tissue

(a)

Hair shaft

Sweat pores

Sebaceous (oil) gland

Hair follicle

Hair root

Duct of sudoriferous (sweat) gland

The Landscape of the Skin
Figure 13.2a

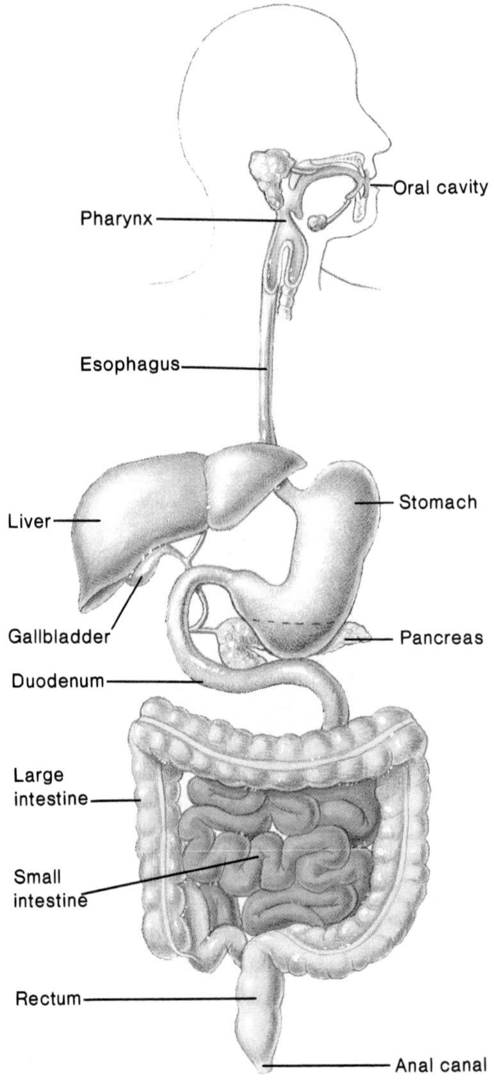

Oral cavity

Pharynx

Esophagus

Liver

Stomach

Gallbladder

Pancreas

Duodenum

Large intestine

Small intestine

Rectum

Anal canal

Areas of the Alimentary Tract
Figure 13.4

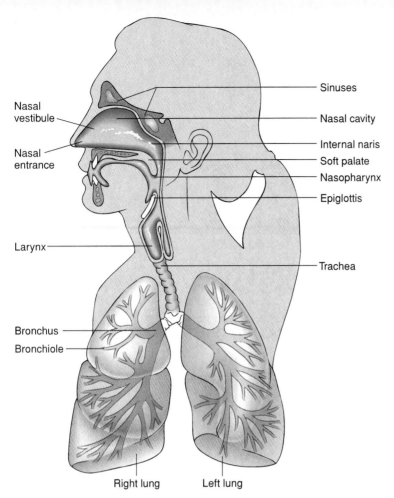

Nasal vestibule

Nasal entrance

Larynx

Bronchus

Bronchiole

Sinuses

Nasal cavity

Internal naris

Soft palate

Nasopharynx

Epiglottis

Trachea

Right lung

Left lung

Colonized Regions of the Respiratory Tract
Figure 13.6

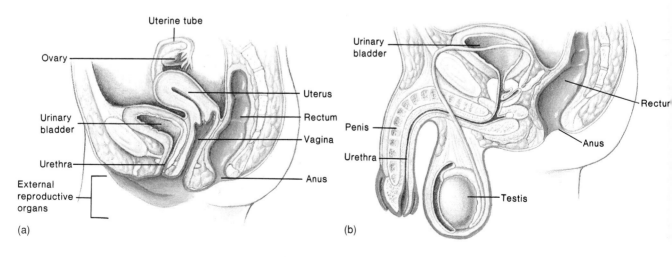

Uterine tube

Ovary

Urinary bladder

Urethra

External reproductive organs

Uterus

Rectum

Vagina

Anus

(a)

Urinary bladder

Penis

Urethra

Rectum

Anus

Testis

(b)

Location of the Female and Male Genitourinary Flora
Figure 13.7

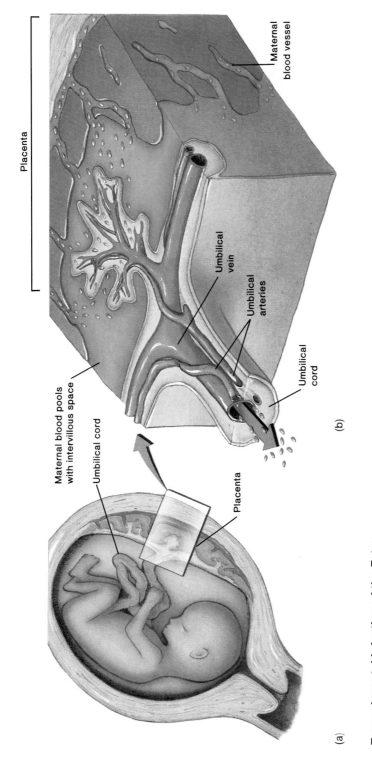

Maternal blood vessel

Placenta

Umbilical vein

Umbilical arteries

Umbilical cord

(b)

Maternal blood pools with intervillous space

Umbilical cord

Placenta

(a)

Transplacental Infection of the Fetus
Figure 13.10

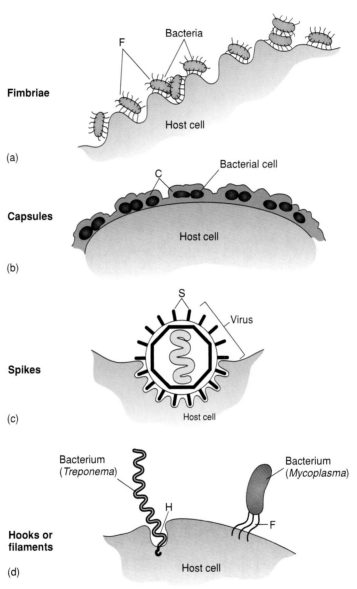

Fimbriae

F

Bacteria

Host cell

(a)

Capsules

C

Bacterial cell

Host cell

(b)

Spikes

S

Virus

Host cell

(c)

Hooks or filaments

Bacterium
(*Treponema*)

H

Bacterium
(*Mycoplasma*)

F

Host cell

(d)

Mechanisms of Adhesion by Pathogens
Figure 13.11

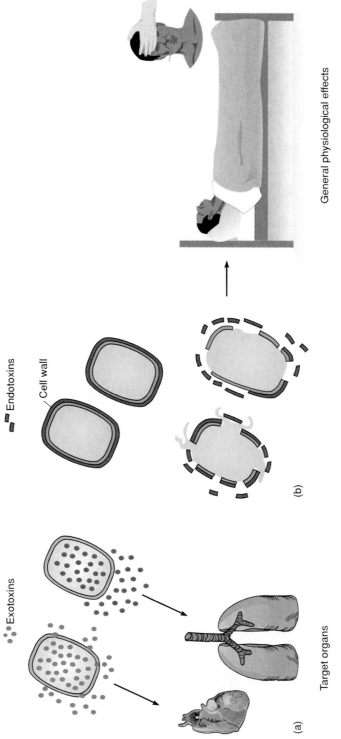

Exotoxins

Endotoxins

Cell wall

(a)

(b)

Target organs

General physiological effects

The Origins and Effects of Circulating Exotoxins and Endotoxins
Figure 13.13

78

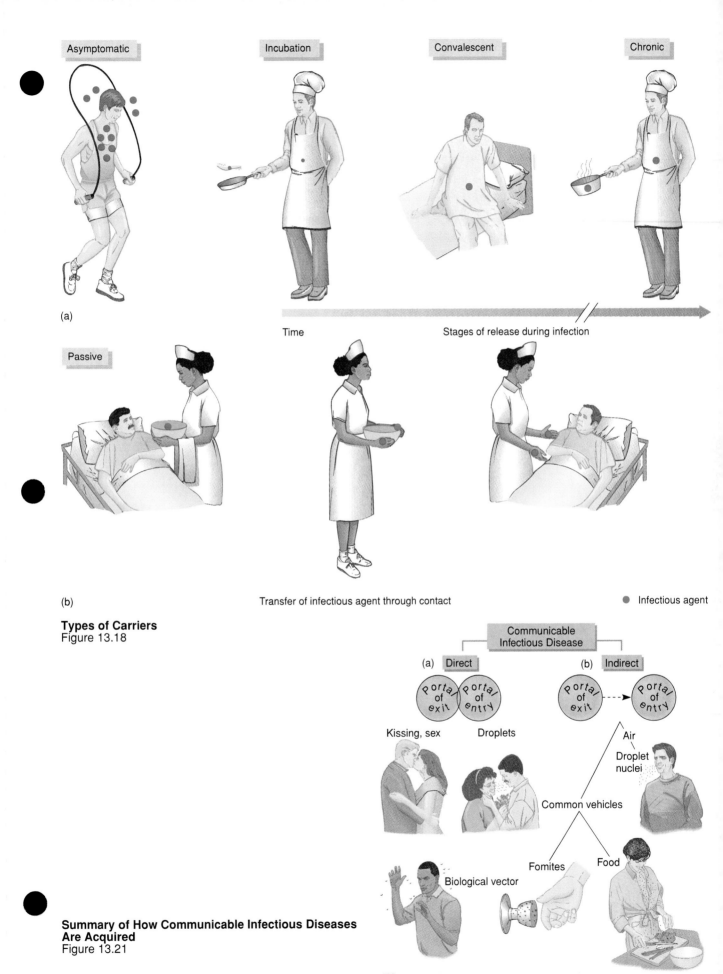

(a)

| Asymptomatic | Incubation | Convalescent | Chronic |

Time → Stages of release during infection

Passive

(b)

Transfer of infectious agent through contact   ● Infectious agent

Types of Carriers
Figure 13.18

Communicable
Infectious Disease

(a) Direct   (b) Indirect

Portal of exit — Portal of entry   Portal of exit ----→ Portal of entry

Kissing, sex   Droplets

Air

Droplet nuclei

Common vehicles

Biological vector   Fomites   Food

**Summary of How Communicable Infectious Diseases
Are Acquired**
Figure 13.21

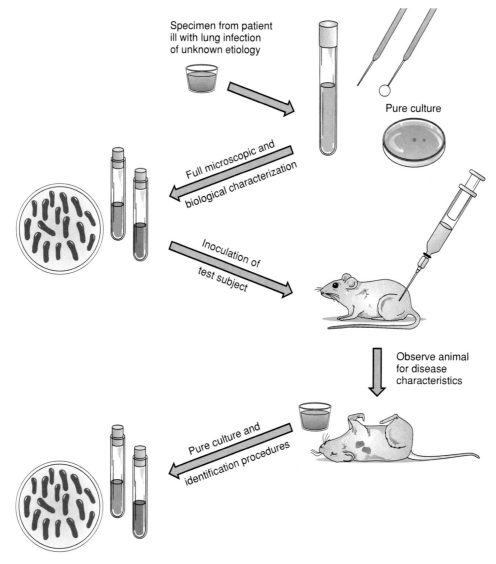

Specimen from patient
ill with lung infection
of unknown etiology

Pure culture

Full microscopic and
biological characterization

Inoculation of
test subject

Observe animal
for disease
characteristics

Pure culture and
identification procedures

Koch's Postulates
Figure 13.24

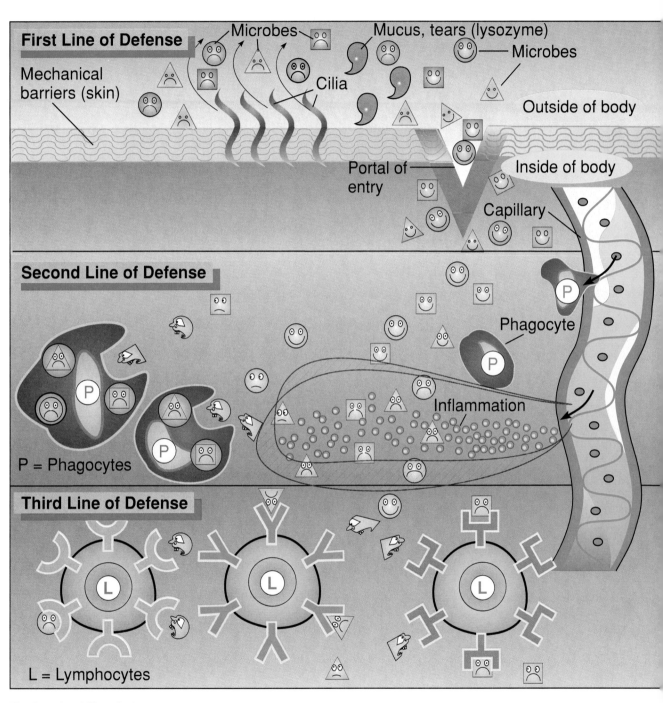

The Levels of Host Defense
Figure 14.1

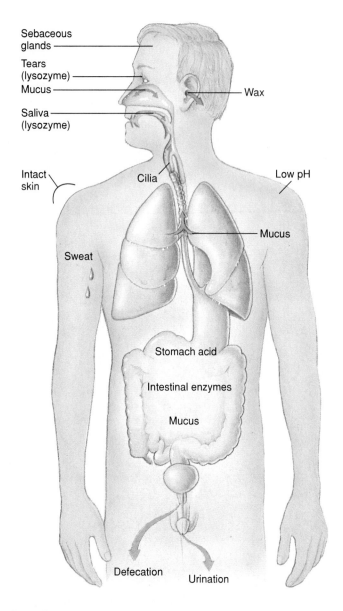

Sebaceous glands

Tears (lysozyme)

Mucus

Saliva (lysozyme)

Wax

Intact skin

Cilia

Low pH

Mucus

Sweat

Stomach acid

Intestinal enzymes

Mucus

Defecation

Urination

The Primary Physical and Chemical Defense Barriers
Figure 14.2

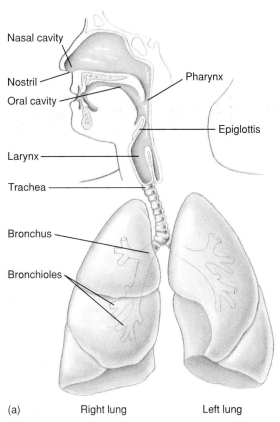

Nasal cavity

Nostril

Oral cavity

Larynx

Trachea

Bronchus

Bronchioles

Pharynx

Epiglottis

(a)   Right lung   Left lung

The Ciliary Defense of the Respiratory Tree
Figure 14.3 a

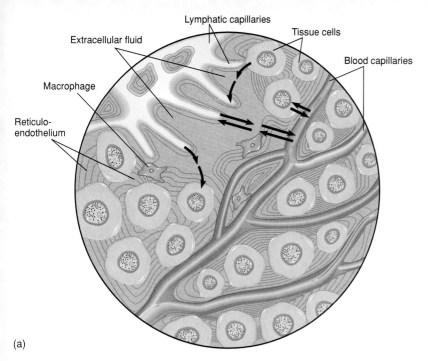

Lymphatic capillaries

Tissue cells

Extracellular fluid

Blood capillaries

Macrophage

Reticulo-
endothelium

(a)

(b)

Blood

Direct connection to veins near heart

Extracellular fluid

Extracellular fluid

Reticuloendothelial

Lymphatics

Extracellular fluid

The Body Compartments Are Separate but Connected
Figure 14.5

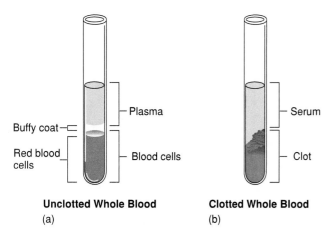

Plasma

Buffy coat

Serum

Red blood
cells

Blood cells

Clot

Unclotted Whole Blood
(a)

Clotted Whole Blood
(b)

The Macroscopic Composition of Whole Blood
Figure 14.7

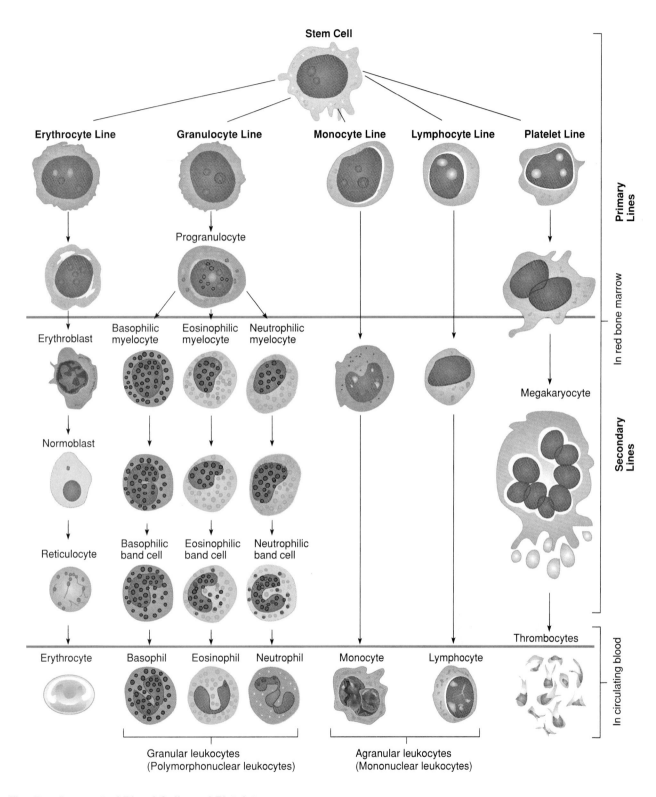

Stem Cell

Erythrocyte Line   Granulocyte Line   Monocyte Line   Lymphocyte Line   Platelet Line

Primary Lines

Progranulocyte

In red bone marrow

Erythroblast   Basophilic myelocyte   Eosinophilic myelocyte   Neutrophilic myelocyte

Megakaryocyte

Normoblast

Secondary Lines

Reticulocyte   Basophilic band cell   Eosinophilic band cell   Neutrophilic band cell

Thrombocytes

Erythrocyte   Basophil   Eosinophil   Neutrophil   Monocyte   Lymphocyte

In circulating blood

Granular leukocytes
(Polymorphonuclear leukocytes)

Agranular leukocytes
(Mononuclear leukocytes)

The Development of Blood Cells and Platelets
Figure 14.9

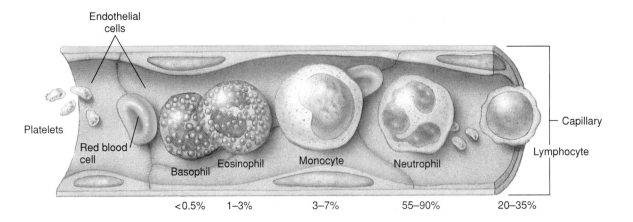

The Microanatomy and Circulating Cells of the Bloodstream
Figure 14.10

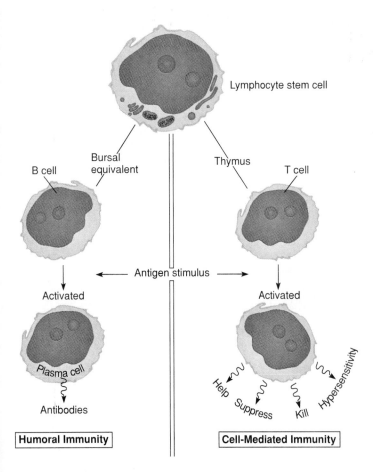

General Development and Functions of Lymphocytes
Figure 14.12

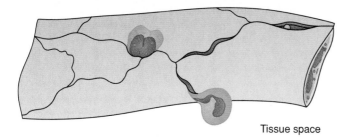

Tissue space

(a)

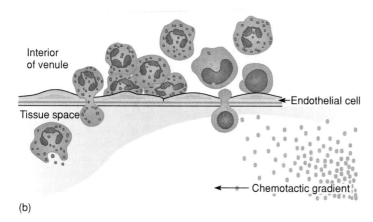

Interior
of venule

Tissue space

Endothelial cell

Chemotactic gradient

(b)

Diapedesis and Chemotaxis of Leukocytes
Figure 14.13

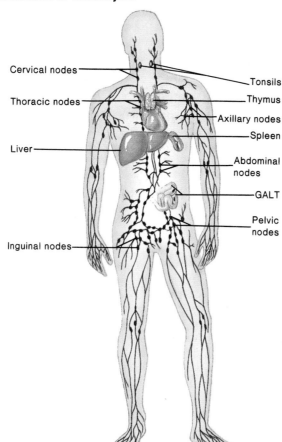

Cervical nodes
Thoracic nodes
Liver
Inguinal nodes

Tonsils
Thymus
Axillary nodes
Spleen
Abdominal nodes
GALT
Pelvic nodes

(a)

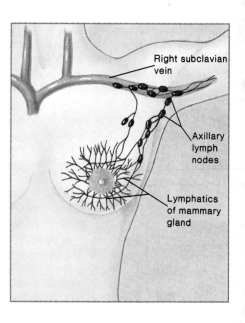

Right subclavian vein
Axillary lymph nodes
Lymphatics of mammary gland

(b)

General Components of the Lymphatic System
Figure 14.14

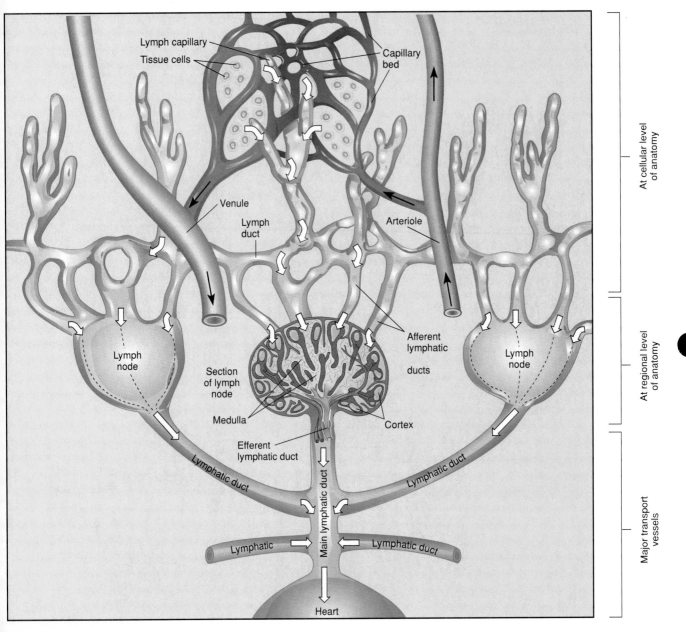

The Circulatory Scheme of the Lymphatic Vessels
Figure 14.15

Labels in figure:
- Lymph capillary
- Tissue cells
- Capillary bed
- Venule
- Lymph duct
- Arteriole
- Lymph node
- Section of lymph node
- Medulla
- Cortex
- Afferent lymphatic ducts
- Lymph node
- Efferent lymphatic duct
- Lymphatic duct
- Main lymphatic duct
- Lymphatic duct
- Lymphatic
- Lymphatic duct
- Heart
- At cellular level of anatomy
- At regional level of anatomy
- Major transport vessels

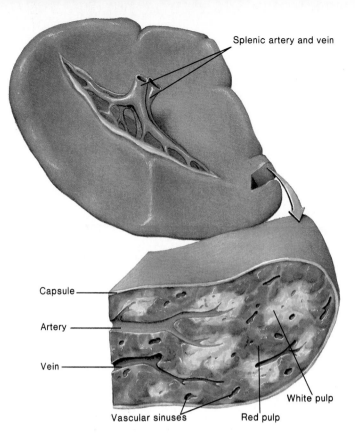

Splenic artery and vein

Capsule

Artery

Vein

Vascular sinuses

Red pulp

White pulp

The Anatomy of the Spleen
Figure 14.16

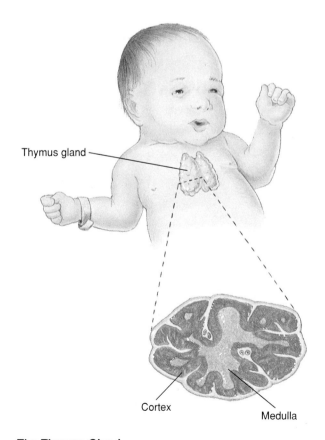

Thymus gland

Cortex

Medulla

The Thymus Gland
Figure 14.17

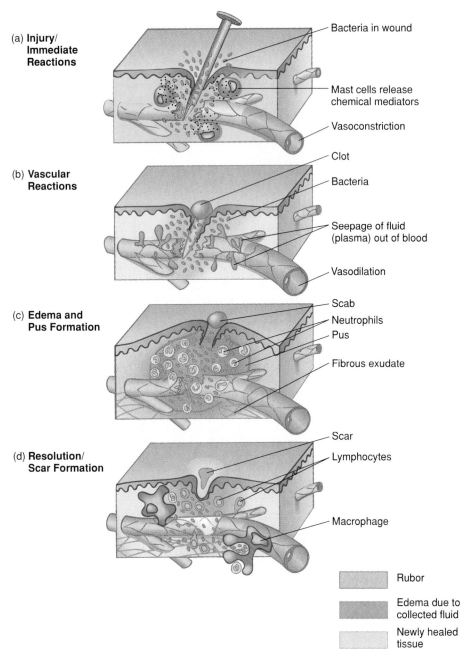

(a) **Injury/
Immediate
Reactions**

Bacteria in wound

Mast cells release
chemical mediators

Vasoconstriction

(b) **Vascular
Reactions**

Clot

Bacteria

Seepage of fluid
(plasma) out of blood

Vasodilation

(c) **Edema and
Pus Formation**

Scab
Neutrophils
Pus

Fibrous exudate

(d) **Resolution/
Scar Formation**

Scar
Lymphocytes

Macrophage

Rubor

Edema due to
collected fluid

Newly healed
tissue

The Major Events in Inflammation
Figure 14.18

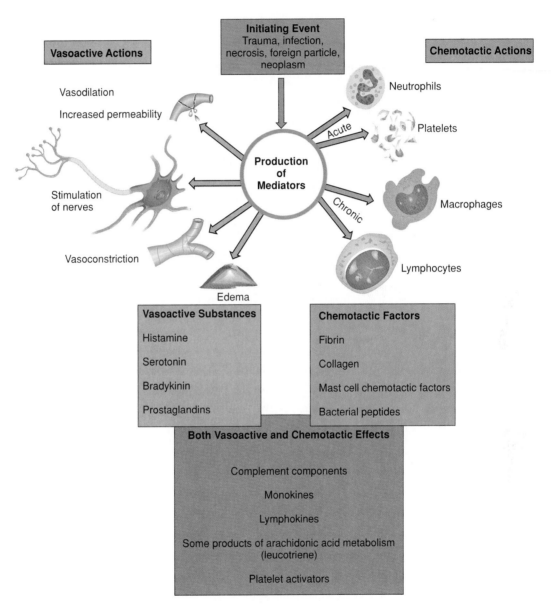

Initiating Event
Trauma, infection, necrosis, foreign particle, neoplasm

Vasoactive Actions

Chemotactic Actions

Vasodilation

Increased permeability

Neutrophils

Acute

Platelets

Production of Mediators

Stimulation of nerves

Macrophages

Chronic

Vasoconstriction

Lymphocytes

Edema

Vasoactive Substances

Histamine

Serotonin

Bradykinin

Prostaglandins

Chemotactic Factors

Fibrin

Collagen

Mast cell chemotactic factors

Bacterial peptides

Both Vasoactive and Chemotactic Effects

Complement components

Monokines

Lymphokines

Some products of arachidonic acid metabolism (leucotriene)

Platelet activators

Chemical Mediators of the Inflammatory Response
Figure 14.19

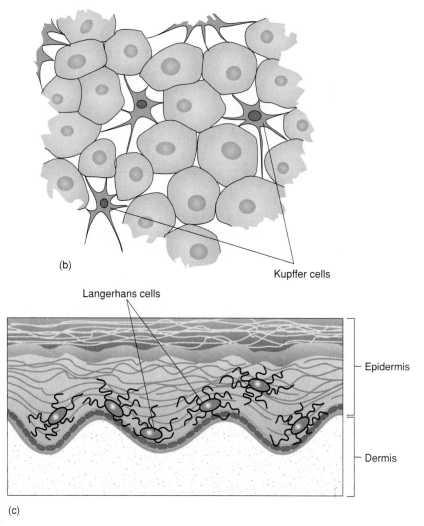

(b)

Kupffer cells

Langerhans cells

Epidermis

Dermis

(c)

Sites Containing Macrophages
Figure 14.21bc

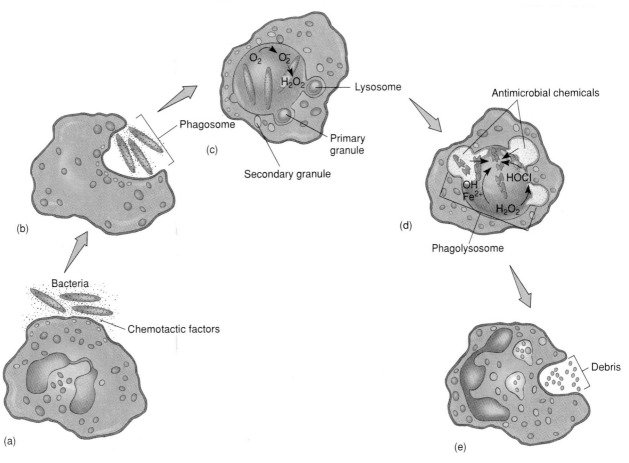

(a)

Bacteria

Chemotactic factors

(b)

Phagosome

(c)

O$_2$ O$_2^-$
        H$_2$O$_2$

Lysosome

Primary
granule

Secondary granule

(d)

Antimicrobial chemicals

OH$^-$
Fe^{2+} HOCl

H$_2$O$_2$

Phagolysosome

(e)

Debris

The Phases in Phagocytosis
Figure 14.22

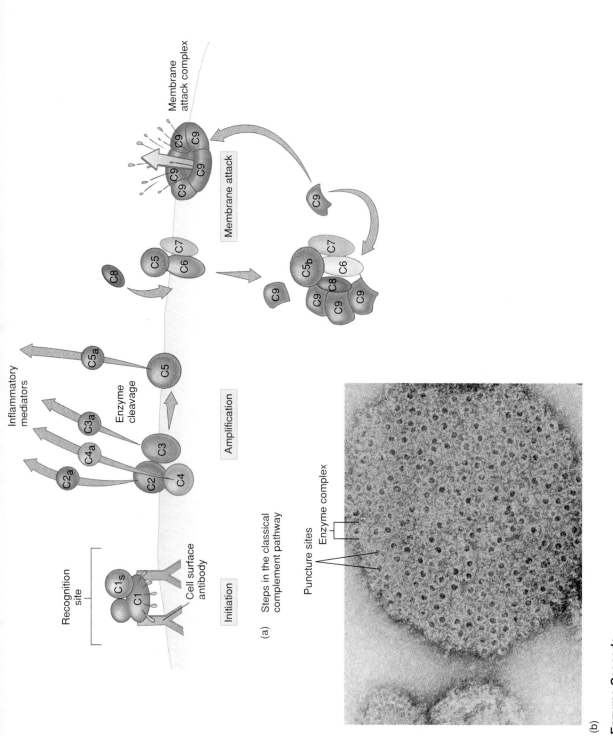

Recognition site

C1s
C1
Cell surface antibody

Initiation

Inflammatory mediators

C2a
C4a
C3a
C5a

C2
C4
C3
C5

Enzyme cleavage

Amplification

C8
C5
C7
C6

C9

C9

C5b
C7
C8
C6
C9
C9
C9
C9

Membrane attack

C9
C9
C9
C9

Membrane attack complex

(a)   Steps in the classical complement pathway

Puncture sites

Enzyme complex

(b)

Enzyme Cascade
Figure 14.24

94

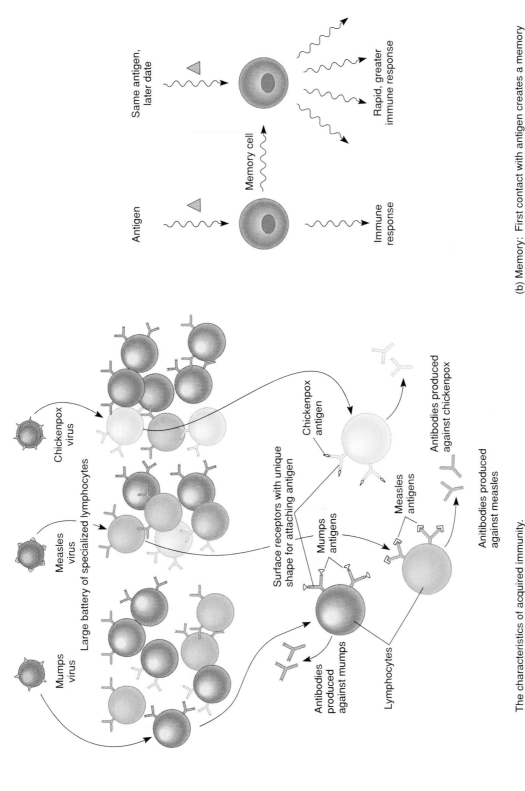

Same antigen, later date

Antigen

Memory cell

Rapid, greater immune response

Immune response

(b) Memory: First contact with antigen creates a memory and quick recall upon second and other future contacts with that antigen.

Chickenpox virus

Measles virus

Mumps virus

Large battery of specialized lymphocytes

Surface receptors with unique shape for attaching antigen

Chickenpox antigen

Mumps antigens

Measles antigens

Antibodies produced against chickenpox

Antibodies produced against measles

Antibodies produced against mumps

Lymphocytes

The characteristics of acquired immunity.
(a) Specificity: Viruses and other infectious agents contain antigen molecules which are specific to a single type of lymphocyte. Binding eventually results in secretion of virus-specific antibodies.

The Characteristics of Acquired Immunity
Figure 14.25

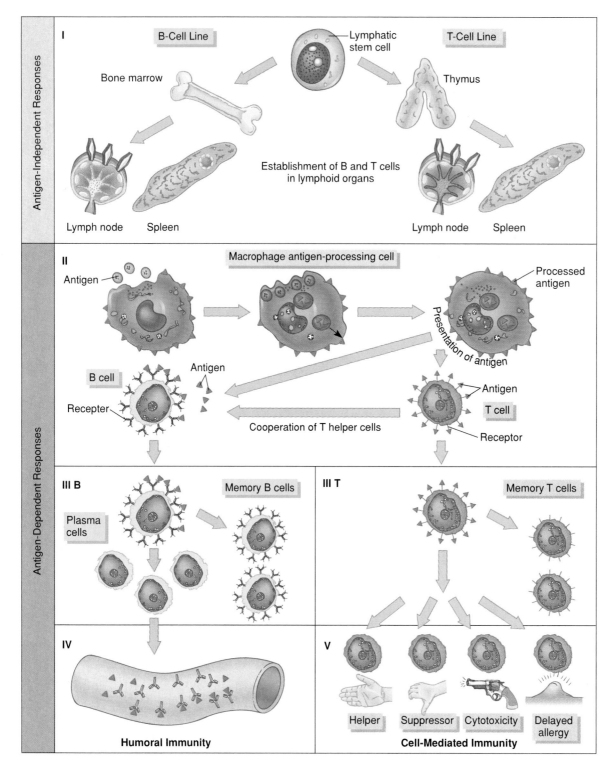

Overview of the Stages of Lymphocyte Development and Function
Figure 15.1

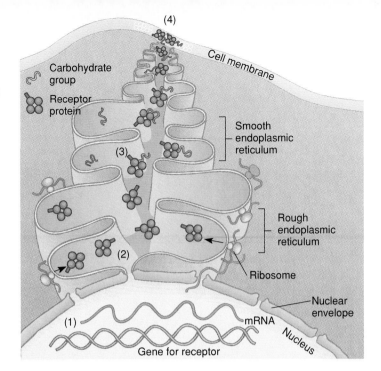

Receptor Formation in a Developing Cell
Figure 15.2

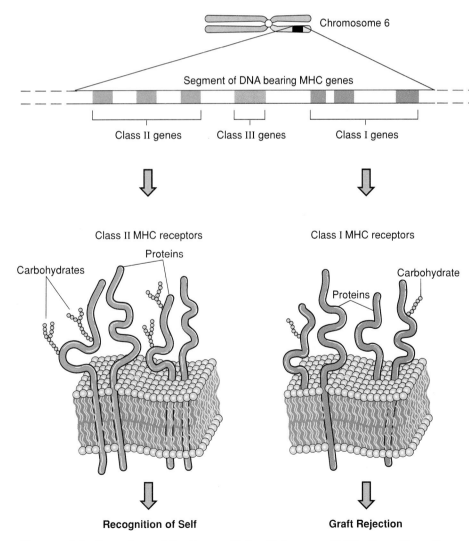

Glycoprotein Receptors of the Human Major Histocompatibility Gene Complex
Figure 15.3

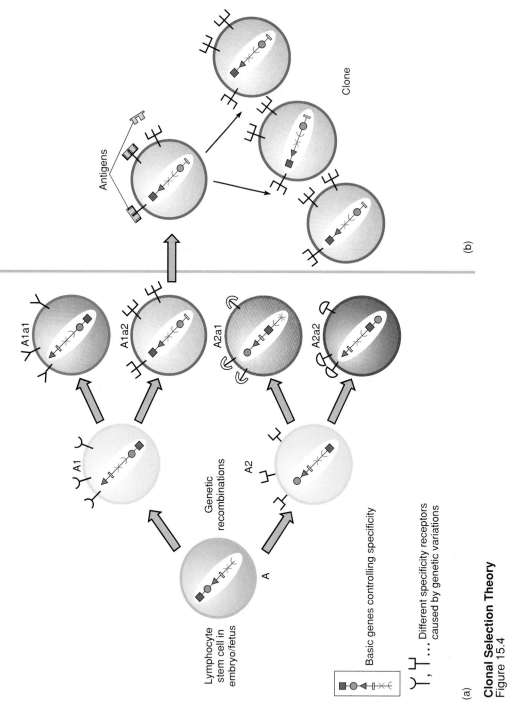

(a)

Lymphocyte stem cell in embryo/fetus

Genetic recombinations

A

A1

A2

A1a1

A1a2

A2a1

A2a2

Antigens

Clone

(b)

Basic genes controlling specificity

Y, T ... Different specificity receptors caused by genetic variations

Clonal Selection Theory
Figure 15.4

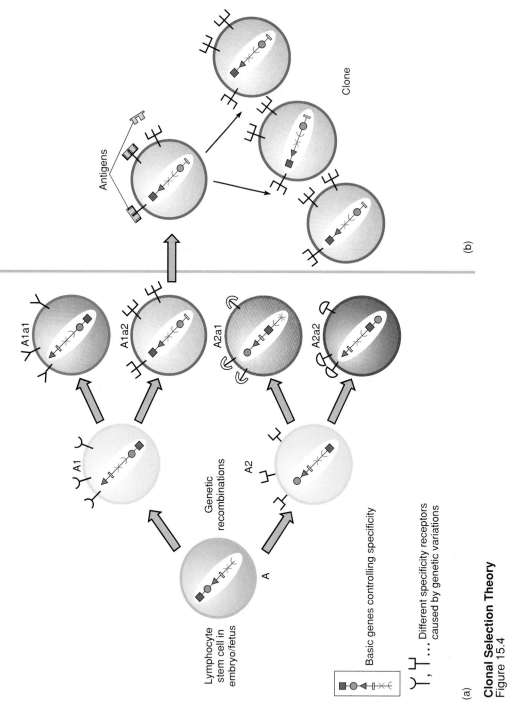

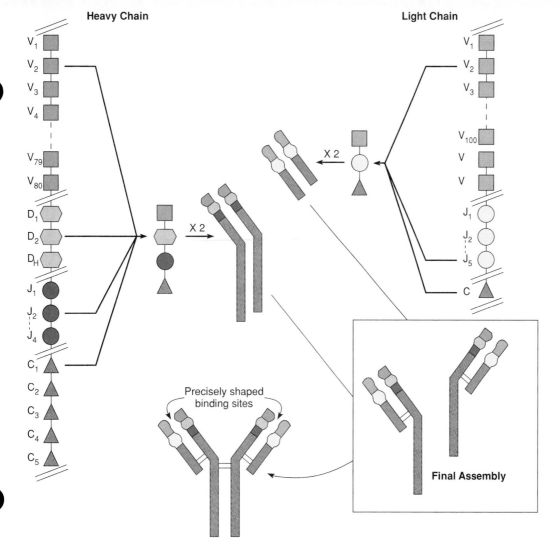

Heavy Chain

V₁ V₂ V₃ V₄ ... V₇₉ V₈₀ D₁ D₂ Dₕ J₁ J₂ J₄ C₁ C₂ C₃ C₄ C₅

Light Chain

V₁ V₂ V₃ ... V₁₀₀ V V J₁ J₂ J₅ C

X 2

X 2

Precisely shaped binding sites

Final Assembly

A Simplified Look at Immunoglobulin Genetics
Figure 15.6

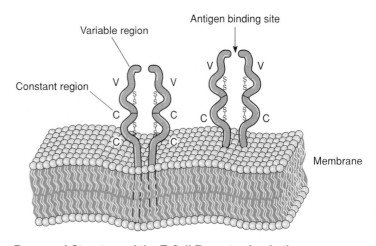

Variable region

Antigen binding site

Constant region

V   V   V   V

C   C   C   C

C   C

Membrane

Proposed Structure of the T-Cell Receptor for Antigen
Figure 15.7

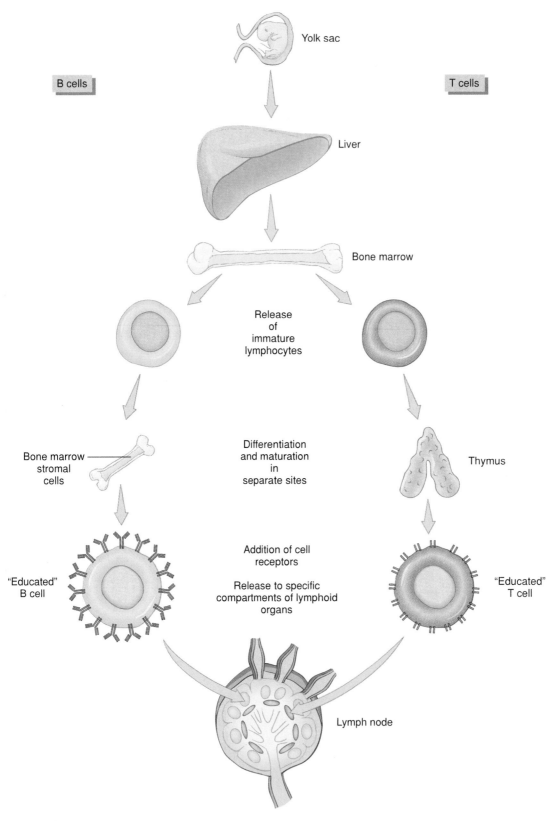

Yolk sac

B cells   T cells

Liver

Bone marrow

Release
of
immature
lymphocytes

Bone marrow
stromal
cells

Differentiation
and maturation
in
separate sites

Thymus

Addition of cell
receptors

Release to specific
compartments of lymphoid
organs

"Educated"
B cell

"Educated"
T cell

Lymph node

Major Stages in the Development of B and T Cells
Figure 15.8

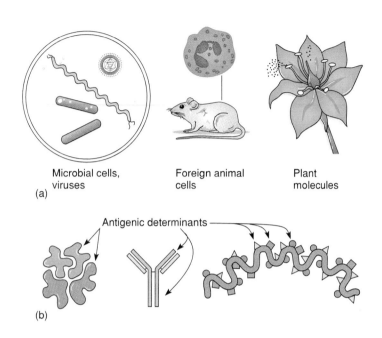

(a) Microbial cells, viruses | Foreign animal cells | Plant molecules

Antigenic determinants

(b)

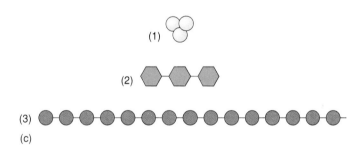

(1)

(2)

(3)

(c)

Characteristics of Antigens
Figure 15.9

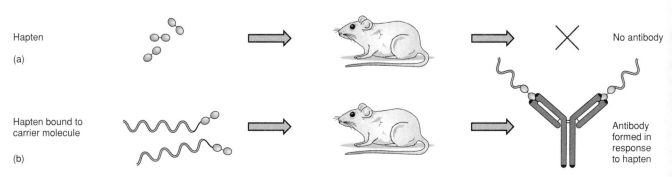

Hapten
(a)

No antibody

Hapten bound to carrier molecule

(b)

Antibody formed in response to hapten

The Hapten-Carrier Phenomenon
Figure 15.11

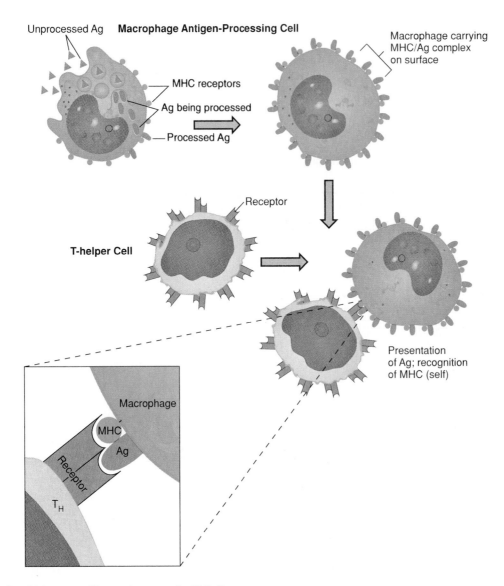

Unprocessed Ag

Macrophage Antigen-Processing Cell

MHC receptors

Ag being processed

Processed Ag

Macrophage carrying
MHC/Ag complex
on surface

Receptor

T-helper Cell

Presentation
of Ag; recognition
of MHC (self)

Macrophage

MHC

Ag

Receptor

T$_H$

Cell Cooperation between a Macrophage and a T-Cell
Figure 15.12

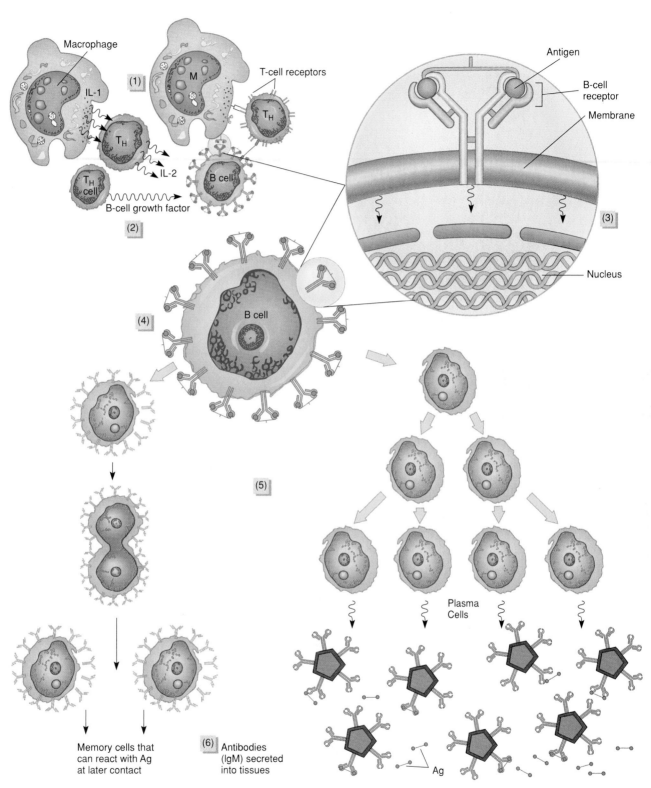

Macrophage

IL-1

(1)

M

T-cell receptors

T_H

T_H

IL-2

T_H cell

B cell

B-cell growth factor

(2)

Antigen

B-cell receptor

Membrane

(3)

Nucleus

(4) B cell

B cell

(5)

Plasma Cells

Memory cells that can react with Ag at later contact

(6) Antibodies (IgM) secreted into tissues

Ag

Events in B-Cell Activation
Figure 15.13

103

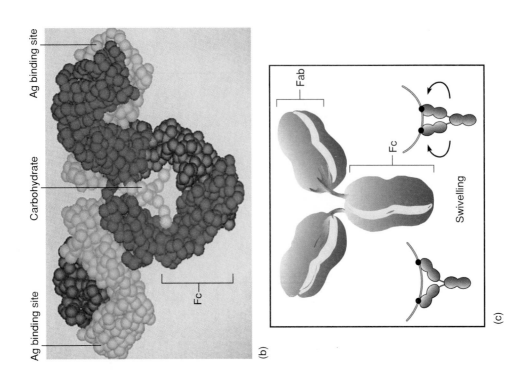

(b)

(c)

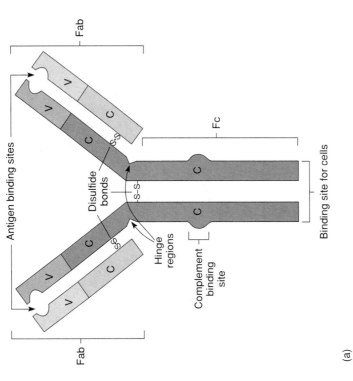

(a)

Working Models of Antibody Structure
Figure 15.14

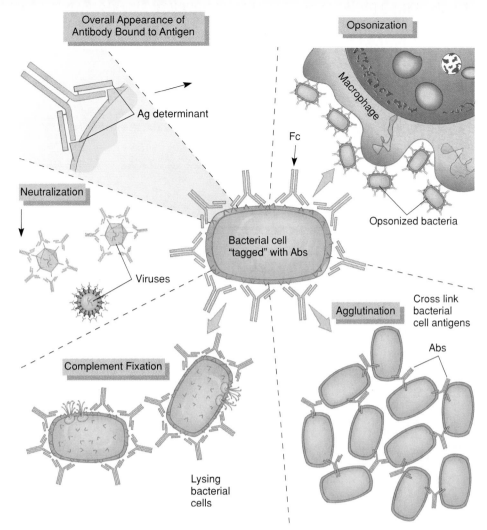

Summary of Antibody Functions
Figure 15.16

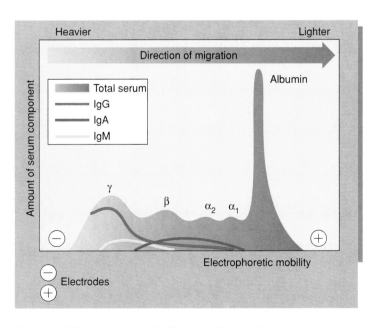

Pattern of Human Serum Following Electrophoresis
Figure 15.17

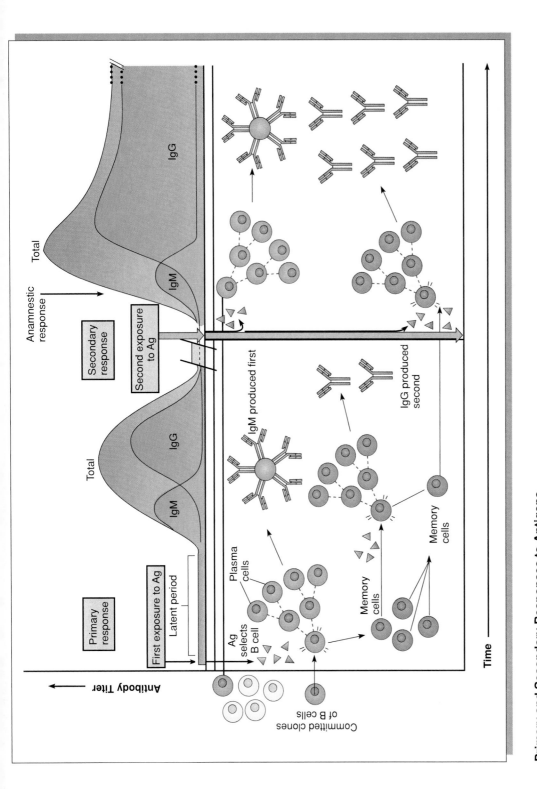

Primary and Secondary Responses to Antigens
Figure 15.18

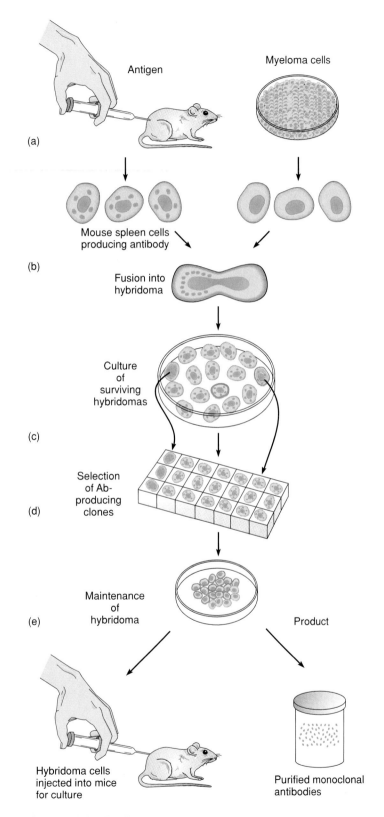

(a)

Antigen

Myeloma cells

Mouse spleen cells
producing antibody

(b)

Fusion into
hybridoma

Culture
of
surviving
hybridomas

(c)

(d)

Selection
of Ab-
producing
clones

(e)

Maintenance
of
hybridoma

Product

Hybridoma cells
injected into mice
for culture

Purified monoclonal
antibodies

Technique for Producing Monoclonal Antibodies
Figure 15.19

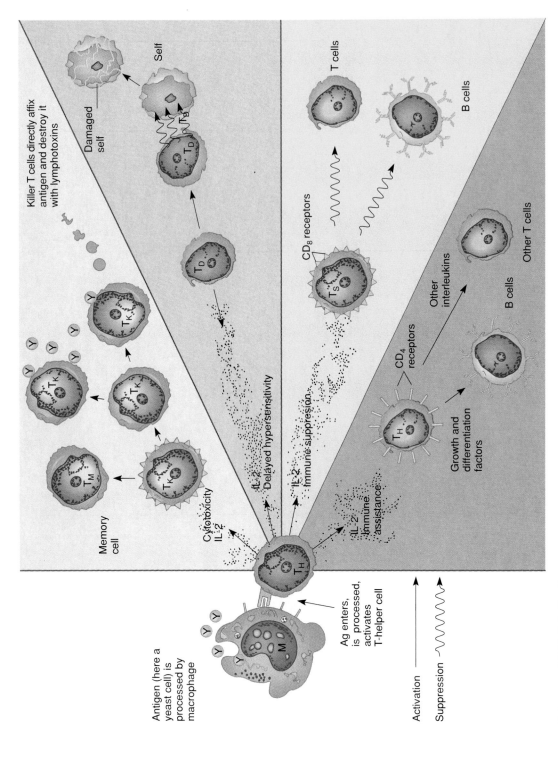

Killer T cells directly affix antigen and destroy it with lymphotoxins

Damaged self

Self

T_B

T_D

T_D

T_K

T_K

T_K

T_M

T_K

Memory cell

Cytotoxicity
IL-2

IL-2
Delayed hypersensitivity

IL-2
Immune suppression

IL-2
Immune assistance

T_H

Antigen (here a yeast cell) is processed by macrophage

M

Ag enters, is processed, activates T-helper cell

T cells

B cells

CD_8 receptors

T_S

CD_4 receptors

T_H

Other interleukins

B cells

Other T cells

Growth and differentiation factors

Activation
Suppression

Scheme of T-Cell Activation
Figure 15.20

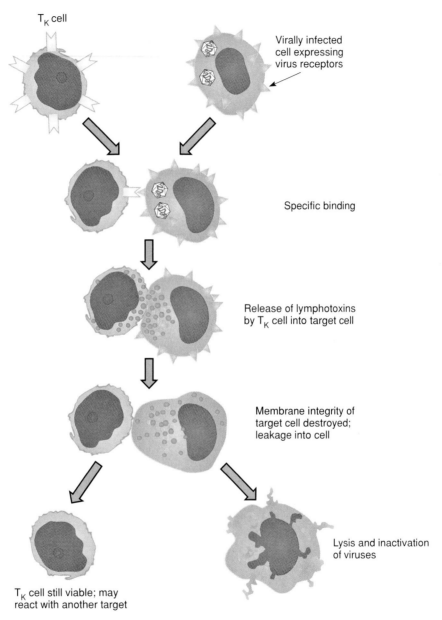

T_K cell

Virally infected cell expressing virus receptors

Specific binding

Release of lymphotoxins by T_K cell into target cell

Membrane integrity of target cell destroyed; leakage into cell

Lysis and inactivation of viruses

T_K cell still viable; may react with another target

Stages of Cell-Mediated Cytotoxicity
Figure 15.21

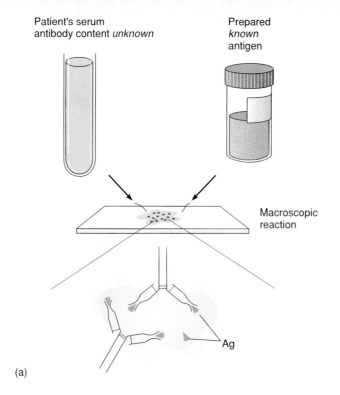

Patient's serum
antibody content *unknown*

Prepared
known
antigen

Macroscopic
reaction

Ag

(a)

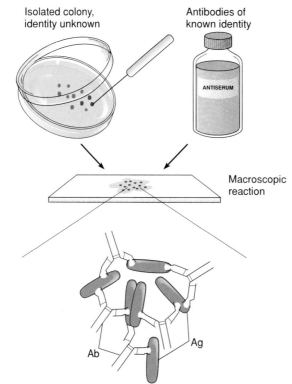

Isolated colony,
identity unknown

Antibodies of
known identity

ANTISERUM

Macroscopic
reaction

Ab

Ag

(b)

Basic Principles of Testing Using Antibodies and Antigens
Figure 16.3

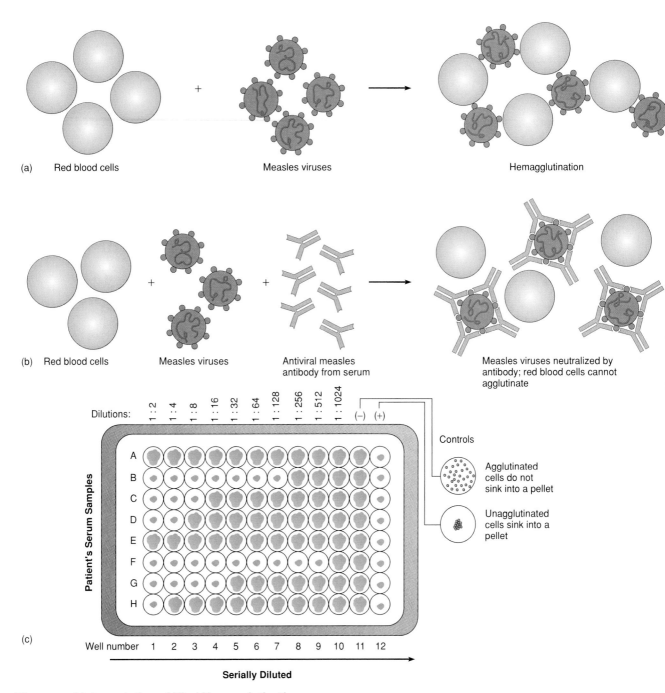

(a) Red blood cells   Measles viruses   Hemagglutination

(b) Red blood cells   Measles viruses   Antiviral measles antibody from serum   Measles viruses neutralized by antibody; red blood cells cannot agglutinate

Dilutions: 1 : 2   1 : 4   1 : 8   1 : 16   1 : 32   1 : 64   1 : 128   1 : 256   1 : 512   1 : 1024   (−)   (+)

Controls

Agglutinated cells do not sink into a pellet

Unagglutinated cells sink into a pellet

Patient's Serum Samples

A B C D E F G H

(c)

Well number 1 2 3 4 5 6 7 8 9 10 11 12

Serially Diluted

Theory and Interpretation of Viral Hemagglutination
Figure 16.6

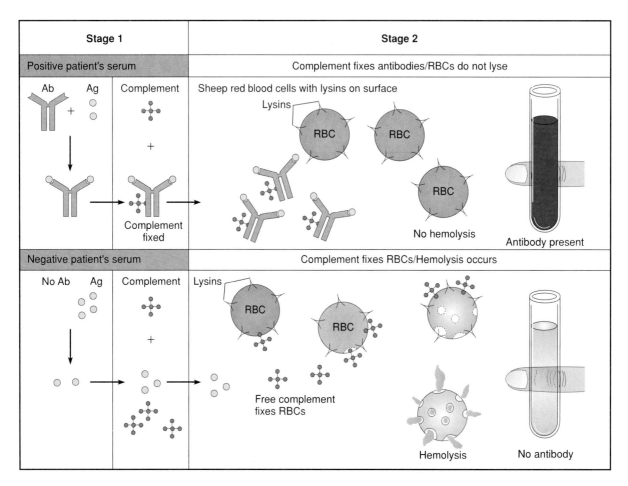

Stage 1	Stage 2

Stage 1 — Positive patient's serum

Ab Ag Complement

Complement fixed

Stage 2 — Complement fixes antibodies/RBCs do not lyse

Sheep red blood cells with lysins on surface

Lysins

RBC   RBC   RBC

No hemolysis

Antibody present

Stage 1 — Negative patient's serum

No Ab Ag Complement

Stage 2 — Complement fixes RBCs/Hemolysis occurs

Lysins

RBC   RBC

Free complement fixes RBCs

Hemolysis

No antibody

Complement Fixation Test
Figure 16.10

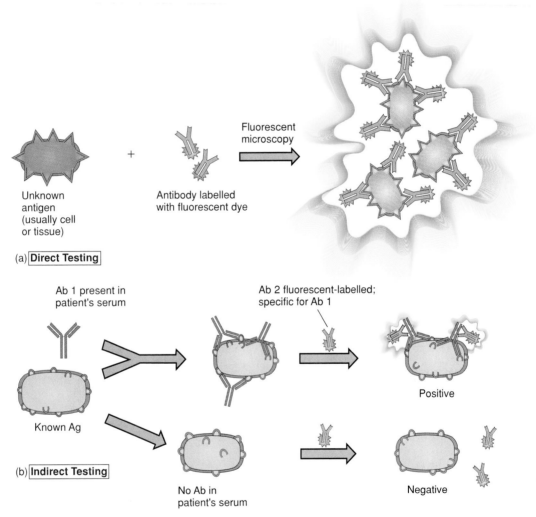

Fluorescent
microscopy

Unknown
antigen
(usually cell
or tissue)

Antibody labelled
with fluorescent dye

(a) Direct Testing

Ab 1 present in
patient's serum

Ab 2 fluorescent-labelled;
specific for Ab 1

Known Ag

Positive

(b) Indirect Testing

No Ab in
patient's serum

Negative

Immunofluorescence Testing
Figure 16.11

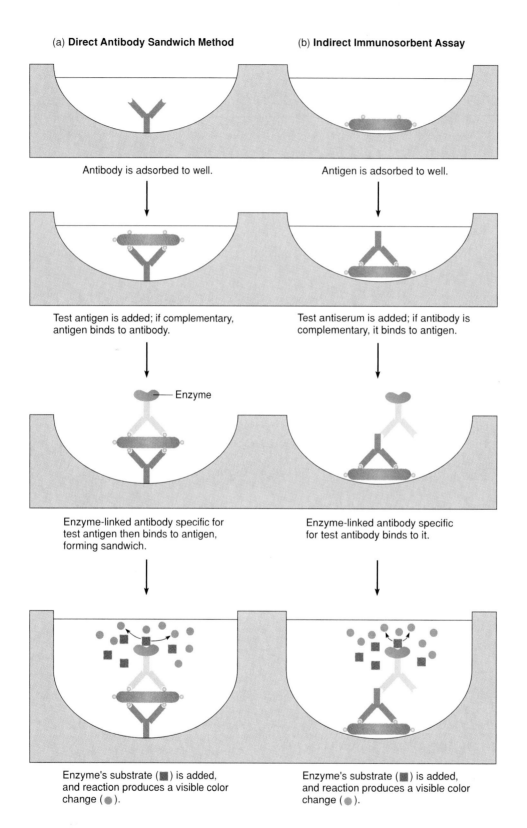

(a) **Direct Antibody Sandwich Method**

Antibody is adsorbed to well.

Test antigen is added; if complementary, antigen binds to antibody.

Enzyme

Enzyme-linked antibody specific for test antigen then binds to antigen, forming sandwich.

Enzyme's substrate (■) is added, and reaction produces a visible color change (●).

(b) **Indirect Immunosorbent Assay**

Antigen is adsorbed to well.

Test antiserum is added; if antibody is complementary, it binds to antigen.

Enzyme-linked antibody specific for test antibody binds to it.

Enzyme's substrate (■) is added, and reaction produces a visible color change (●).

Methods of ELISA Testing
Figure 16.12

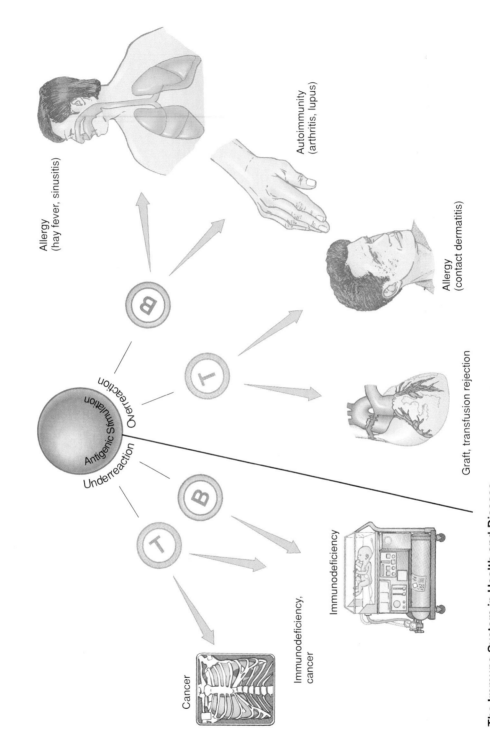

Allergy
(hay fever, sinusitis)

Autoimmunity
(arthritis, lupus)

Allergy
(contact dermatitis)

Graft, transfusion rejection

Antigenic Stimulation

Overreaction

Underreaction

Cancer

Immunodeficiency, cancer

Immunodeficiency

The Immune System in Health and Disease
Figure 17.1

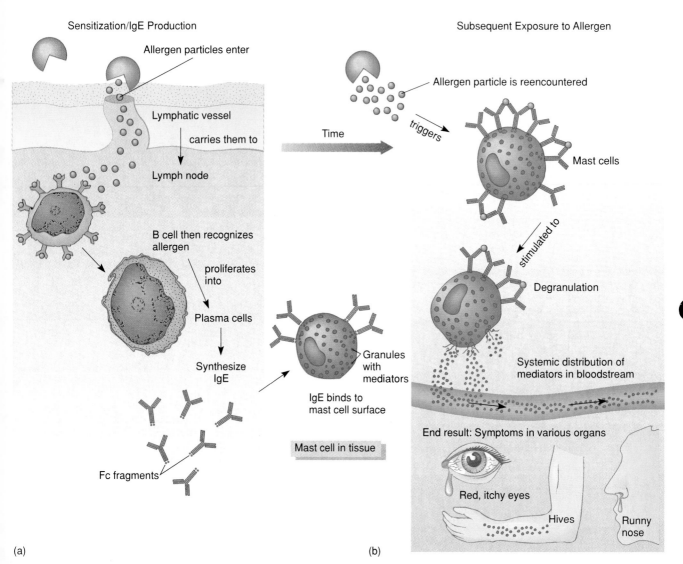

Sensitization/IgE Production

Allergen particles enter

Lymphatic vessel

carries them to

Lymph node

B cell then recognizes allergen

proliferates into

Plasma cells

Synthesize IgE

Fc fragments

Time

Subsequent Exposure to Allergen

Allergen particle is reencountered

triggers

Mast cells

stimulated to

Degranulation

Systemic distribution of mediators in bloodstream

End result: Symptoms in various organs

Red, itchy eyes

Hives

Runny nose

Granules with mediators

IgE binds to mast cell surface

Mast cell in tissue

(a)

(b)

Schematic View of Cellular Reactions during the Allergic Response
Figure 17.3

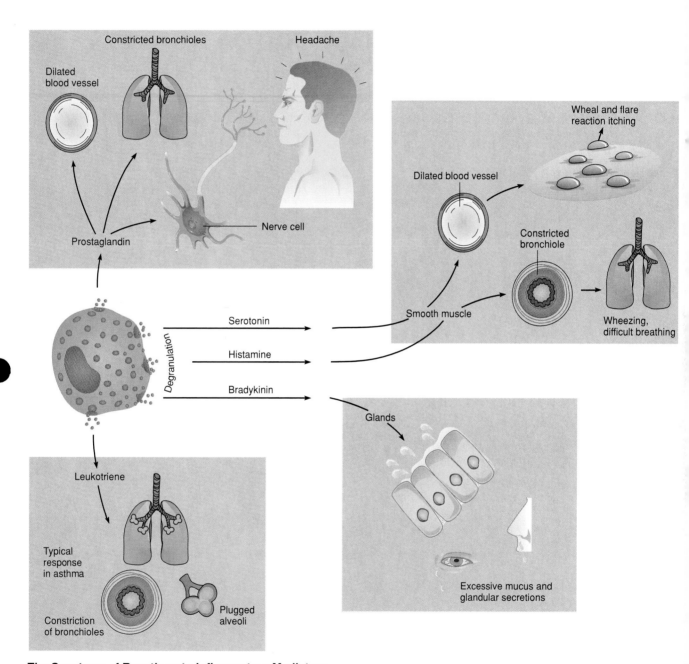

The Spectrum of Reactions to Inflammatory Mediators
Figure 17.4

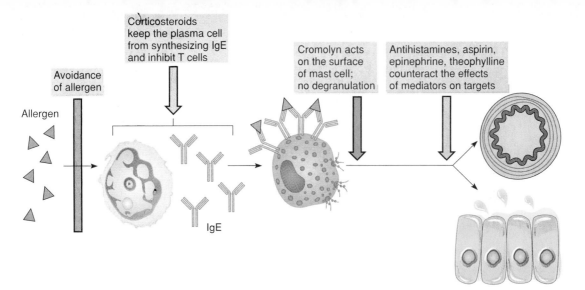

Strategies for Circumventing Allergic Attacks
Figure 17.7

Avoidance of allergen

Corticosteroids keep the plasma cell from synthesizing IgE and inhibit T cells

Cromolyn acts on the surface of mast cell; no degranulation

Antihistamines, aspirin, epinephrine, theophylline counteract the effects of mediators on targets

Allergen

IgE

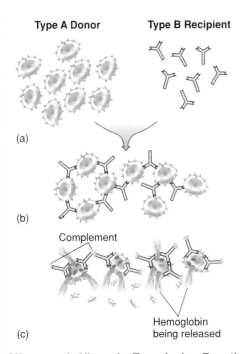

Type A Donor   Type B Recipient

(a)

(b)

Complement

Hemoglobin being released

(c)

Microscopic View of a Transfusion Reaction
Figure 17.10

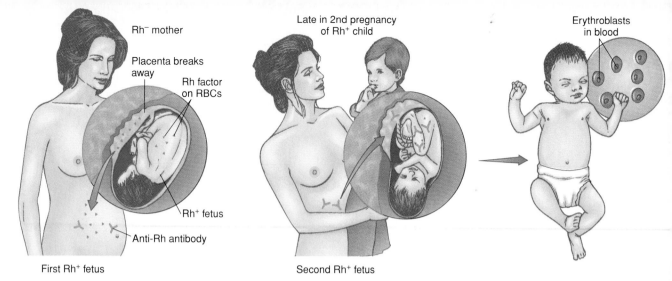

Rh⁻ mother

Placenta breaks away

Rh factor on RBCs

Rh⁺ fetus

Anti-Rh antibody

First Rh⁺ fetus

Late in 2nd pregnancy of Rh⁺ child

Second Rh⁺ fetus

Erythroblasts in blood

The Development and Aftermath of Rh Sensitization
Figure 17.11

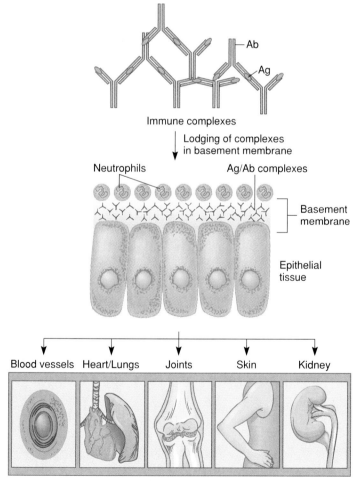

Ab

Ag

Immune complexes

Lodging of complexes in basement membrane

Neutrophils

Ag/Ab complexes

Basement membrane

Epithelial tissue

Blood vessels   Heart/Lungs   Joints   Skin   Kidney

Major organs that can be targets of immune complex deposition

The Background of Immune Complex Disease
Figure 17.13

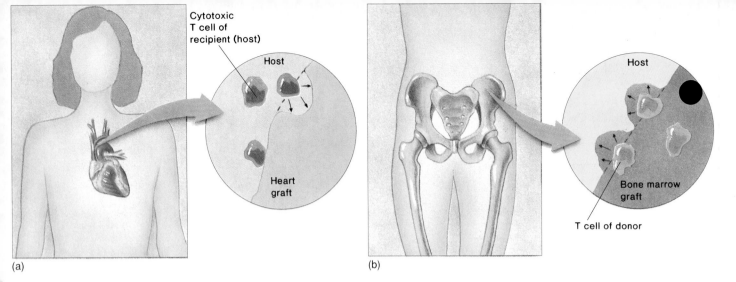

(a)

(b)

Potential Reactions in Transplantation
Figure 17.20

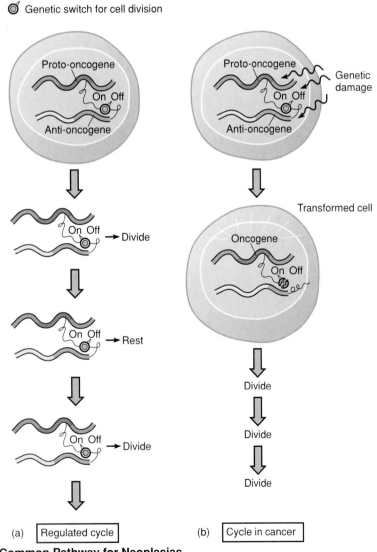

Genetic switch for cell division

Proto-oncogene

On Off

Anti-oncogene

Proto-oncogene

Genetic damage

On Off

Anti-oncogene

On Off → Divide

Transformed cell

Oncogene

On Off

On Off → Rest

Divide

On Off → Divide

Divide

Divide

(a) | Regulated cycle |

(b) | Cycle in cancer |

Common Pathway for Neoplasias
Figure 17.24

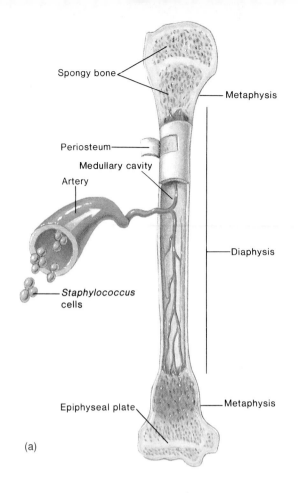

Spongy bone

Metaphysis

Periosteum

Medullary cavity

Artery

Diaphysis

Staphylococcus cells

Metaphysis

Epiphyseal plate

(a)

Staphylococcal Osteomyelitis in a Long Bone
Figure 18.4a

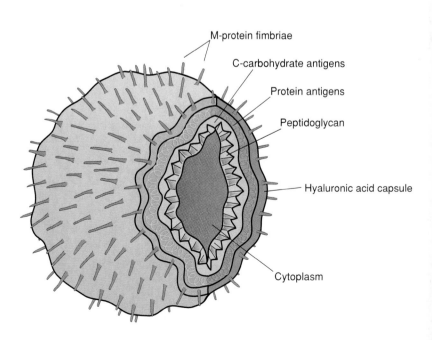

M-protein fimbriae

C-carbohydrate antigens

Protein antigens

Peptidoglycan

Hyaluronic acid capsule

Cytoplasm

Cutaway View of Group A *Streptococcus*
Figure 18.10

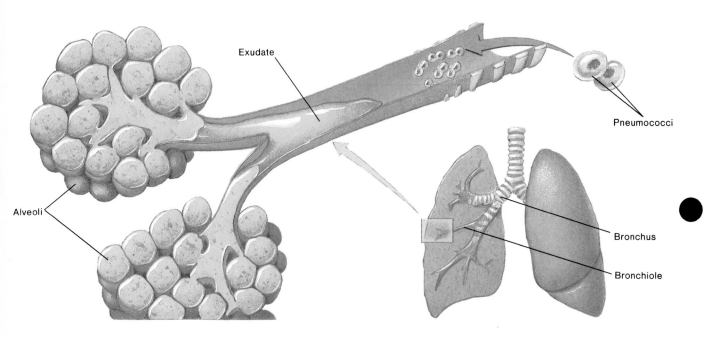

Exudate

Pneumococci

Alveoli

Bronchus

Bronchiole

The Course of Bacterial Pneumonia
Figure 18.18

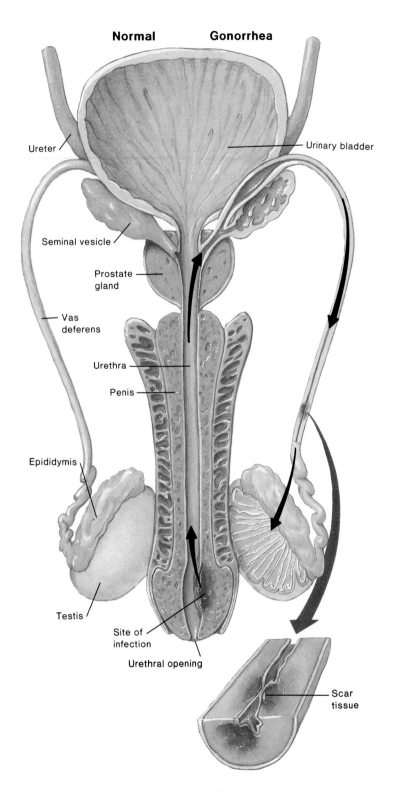

Normal   **Gonorrhea**

Ureter

Urinary bladder

Seminal vesicle

Prostate gland

Vas deferens

Urethra

Penis

Epididymis

Testis

Site of infection

Urethral opening

Scar tissue

Front View of the Male Reproductive Tract
Figure 18.23

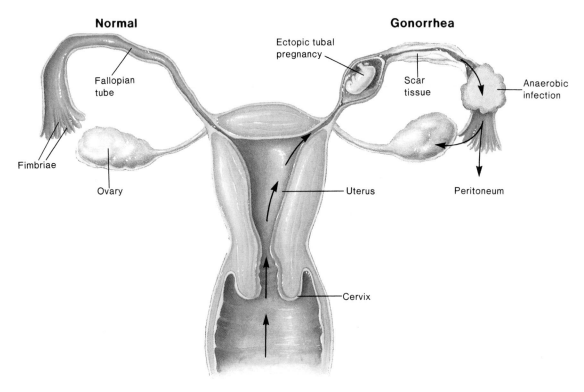

Normal

Fallopian
tube

Fimbriae

Ovary

Gonorrhea

Ectopic tubal
pregnancy

Scar
tissue

Anaerobic
infection

Uterus

Peritoneum

Cervix

Invasive Gonorrhea in Women
Figure 18.24

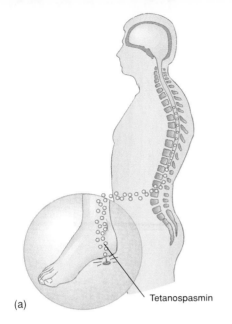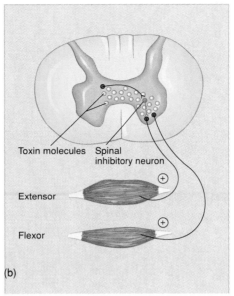

(a)

(b)

Toxin molecules   Spinal
                   inhibitory neuron

Extensor

Flexor

Tetanospasmin

(c)

The Events in Tetanus
Figure 19.6

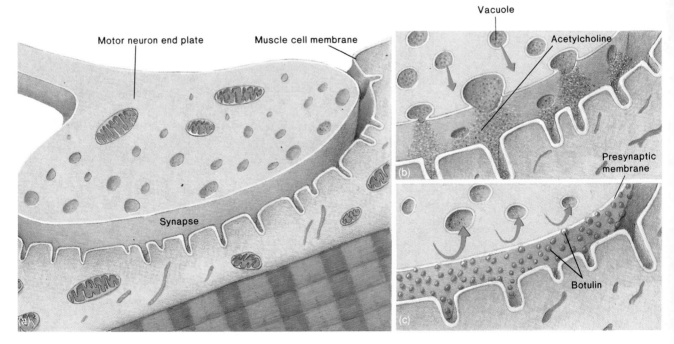

Motor neuron end plate   Muscle cell membrane

Synapse

(a)

Vacuole

Acetylcholine

(b)

Presynaptic
membrane

Botulin

(c)

The Physiological Effects of Botulism Toxin (Botulin)
Figure 19.8

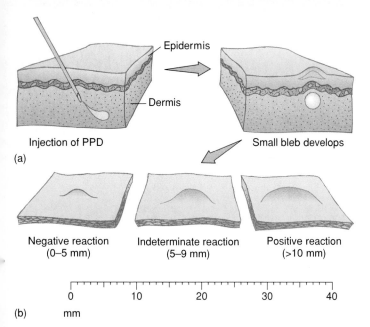

(a)

(b)   mm
0   10   20   30   40

Negative reaction
(0–5 mm)

Indeterminate reaction
(5–9 mm)

Positive reaction
(>10 mm)

Testing for Tuberculosis
Figure 19.18

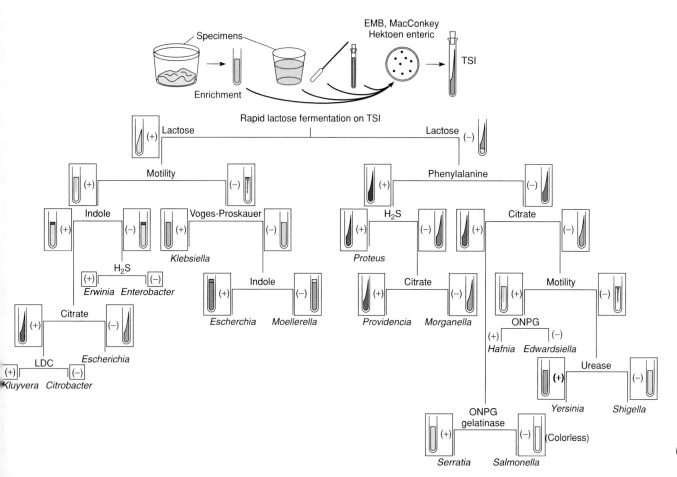

Procedures for Isolating and Identifying Selected Enteric Genera
Figure 20.8

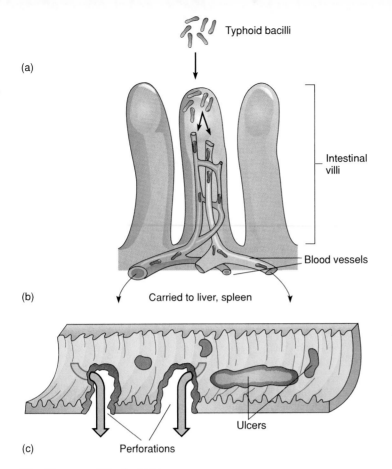

(a)

Typhoid bacilli

Intestinal villi

Blood vessels

(b)

Carried to liver, spleen

Ulcers

(c)

Perforations

The Phases of Typhoid Fever
Figure 20.16

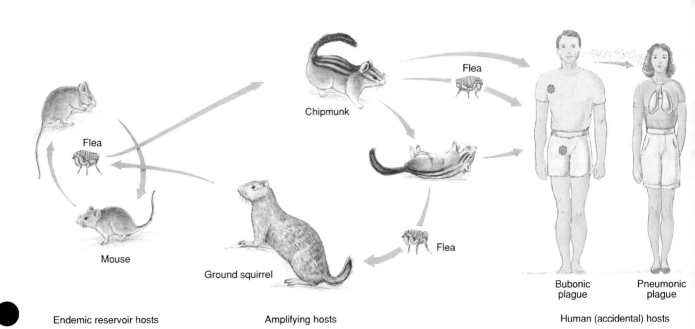

Flea

Chipmunk

Flea

Mouse

Ground squirrel

Flea

Bubonic plague

Pneumonic plague

Endemic reservoir hosts

Amplifying hosts

Human (accidental) hosts

The Infection Cycle of *Y. pestis* Simplified for Clarity
Figure 20.19

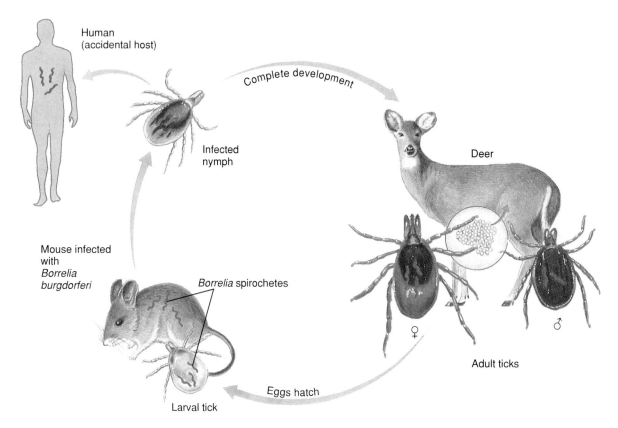

Human
(accidental host)

Complete development

Infected
nymph

Deer

Mouse infected
with
*Borrelia
burgdorferi*

Borrelia spirochetes

♀

♂

Adult ticks

Eggs hatch

Larval tick

The Cycle of Lyme Disease
Figure 21.11

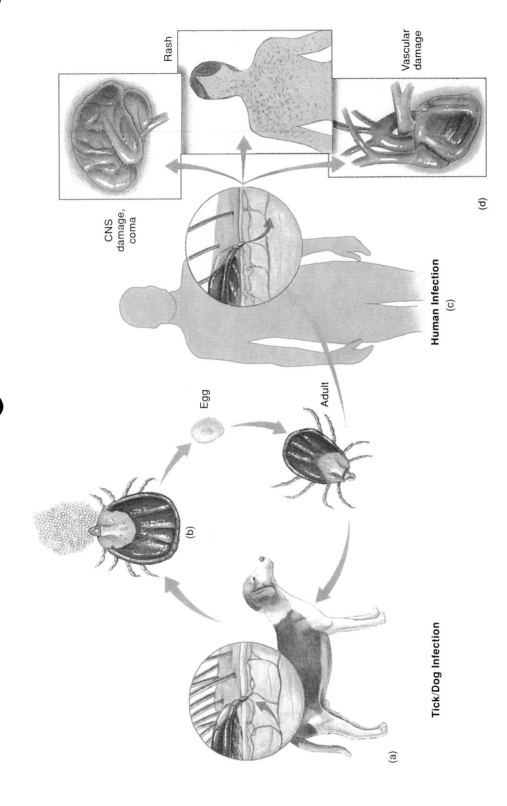

Rash

CNS damage, coma

Vascular damage

Human Infection
(c)

(d)

Egg

Adult

Tick/Dog Infection

(b)

(a)

The Transmission Cycle in Rocky Mountain Spotted Fever
Figure 21.18

129

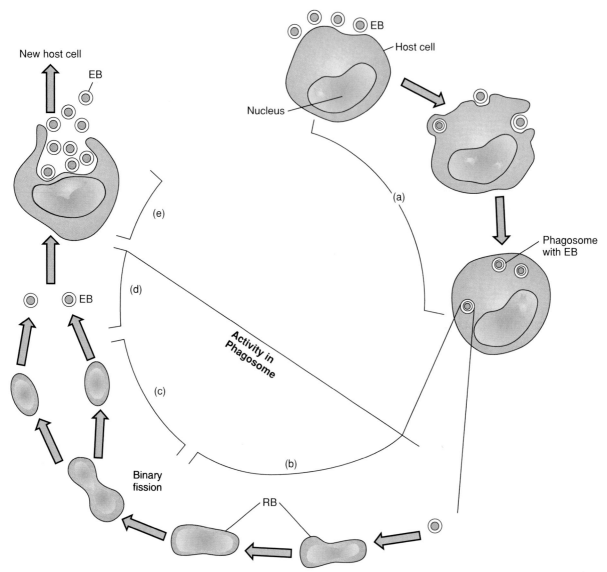

New host cell

EB

Host cell

Nucleus

(a)

Phagosome
with EB

EB

(e)

(d)

Activity in
Phagosome

(c)

(b)

EB

Binary
fission

RB

The Life Cycle of *Chlamydia*
Figure 21.22

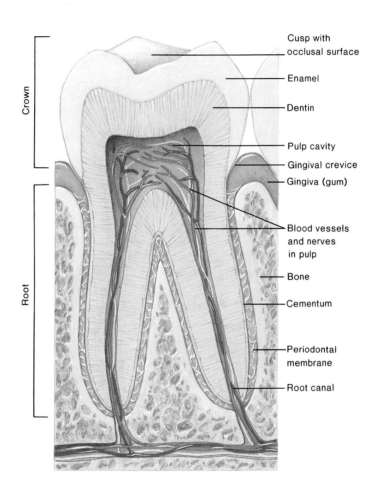

Cusp with occlusal surface

Enamel

Dentin

Pulp cavity

Gingival crevice

Gingiva (gum)

Blood vessels and nerves in pulp

Bone

Cementum

Periodontal membrane

Root canal

Crown

Root

The Anatomy of a Tooth
Figure 21.27

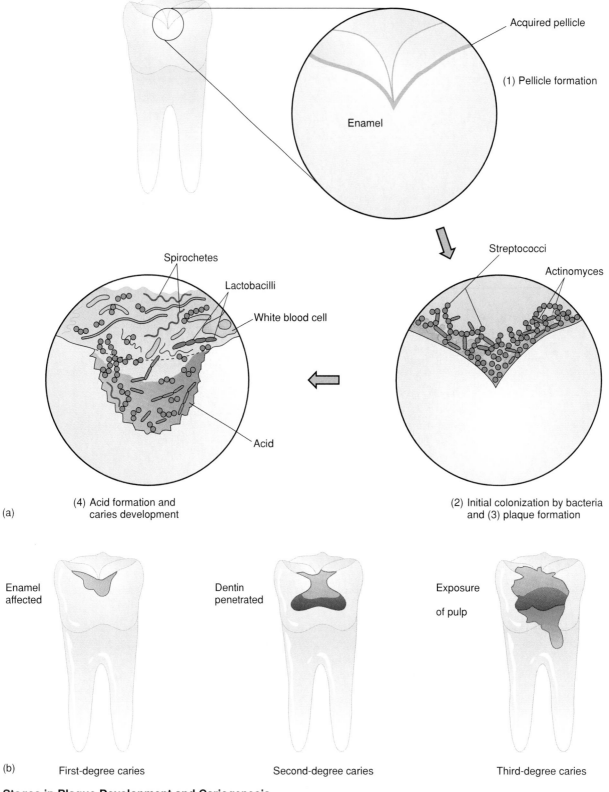

Acquired pellicle

(1) Pellicle formation

Enamel

Streptococci

Actinomyces

(2) Initial colonization by bacteria
and (3) plaque formation

Spirochetes

Lactobacilli

White blood cell

Acid

(4) Acid formation and
caries development

(a)

(b)

Enamel
affected

First-degree caries

Dentin
penetrated

Second-degree caries

Exposure

of pulp

Third-degree caries

Stages in Plaque Development and Cariogenesis
Figure 21.29

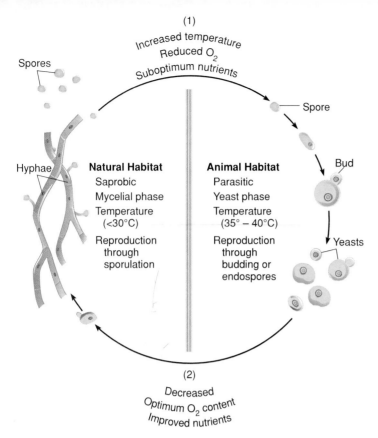

(1)
Increased temperature
Reduced O_2
Suboptimum nutrients

Spores

Spore

Hyphae

Bud

Natural Habitat
Saprobic
Mycelial phase
Temperature (<30°C)
Reproduction through sporulation

Animal Habitat
Parasitic
Yeast phase
Temperature (35° – 40°C)
Reproduction through budding or endospores

Yeasts

(2)
Decreased
Optimum O_2 content
Improved nutrients

The General Changes Associated with Thermal Dimorphism
Figure 22.1

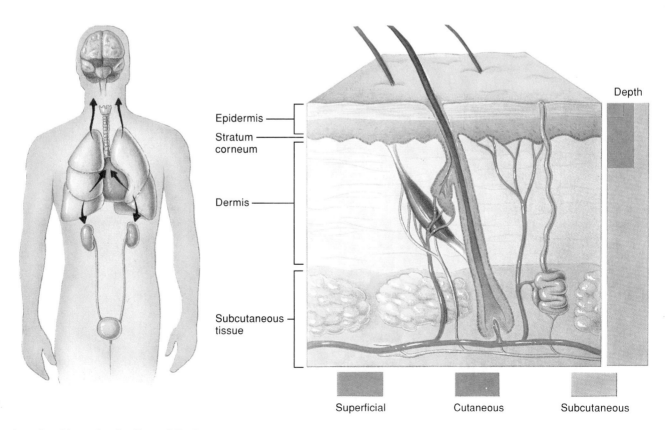

Epidermis
Stratum corneum
Dermis
Subcutaneous tissue

Depth

Superficial   Cutaneous   Subcutaneous

Levels of Invasion by Fungal Pathogens
Figure 22.5

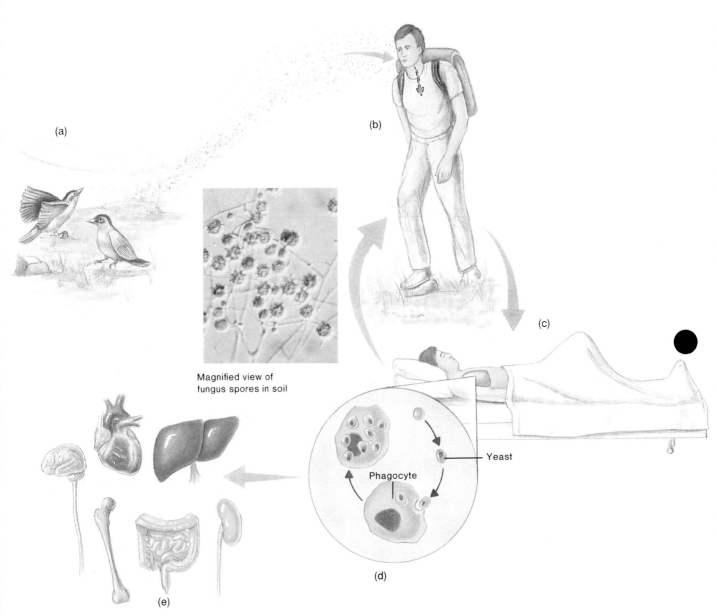

(a)

(b)

Magnified view of
fungus spores in soil

(c)

Yeast

Phagocyte

(d)

(e)

Events in Histoplasmosis
Figure 22.7

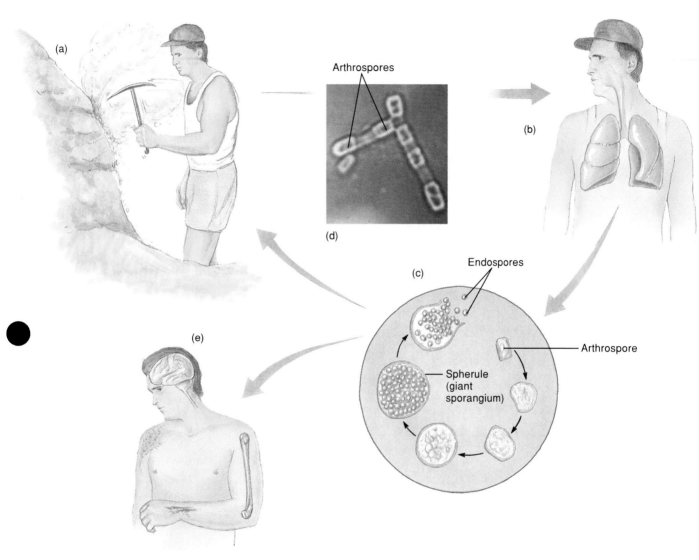

Arthrospores

(a)

(b)

(d)

Endospores

(c)

Arthrospore

Spherule
(giant
sporangium)

(e)

Events in Coccidiodomycosis
Figure 22.8

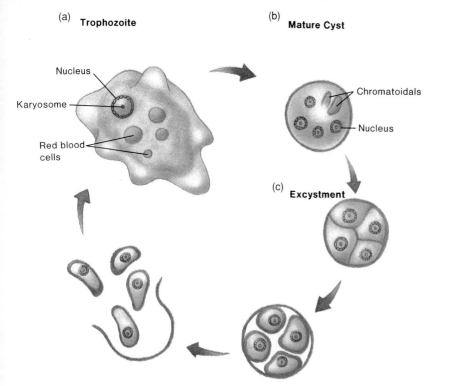

(a) **Trophozoite**

Nucleus

Karyosome

Red blood cells

(b) **Mature Cyst**

Chromatoidals

Nucleus

(c) **Excystment**

Cellular Forms of _Entamoeba histolytica_
Figure 23.1

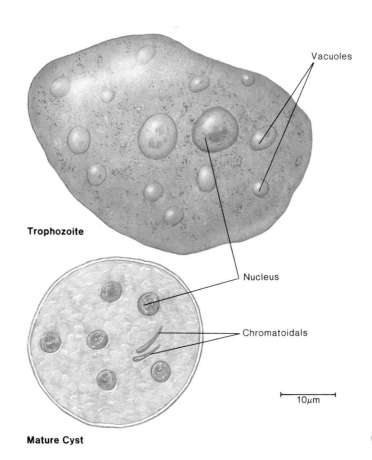

Vacuoles

Trophozoite

Nucleus

Chromatoidals

10μm

Mature Cyst

Entamoeba coli
Figure 23.3

136

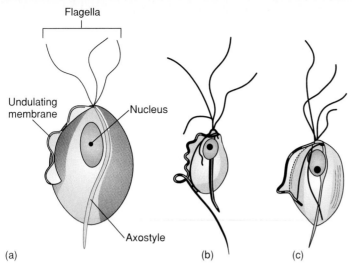

Flagella

Undulating
membrane

Nucleus

Axostyle

(a)

(b)

(c)

The Trichomonads of Humans
Figure 23.6

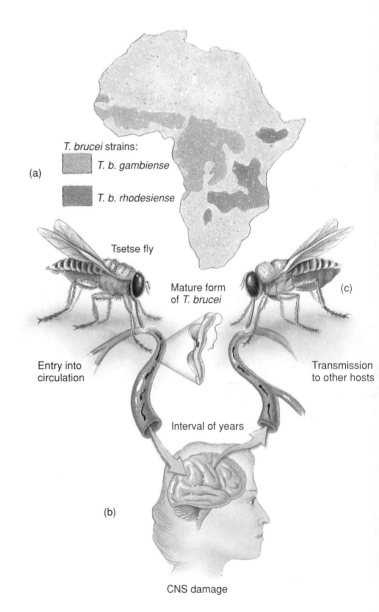

(a)

T. brucei strains:

T. b. gambiense

T. b. rhodesiense

Tsetse fly

Mature form
of *T. brucei*

(c)

Entry into
circulation

Transmission
to other hosts

Interval of years

(b)

CNS damage

Distribution and Generalized Cycle of Trypanosomiasis
Figure 23.8

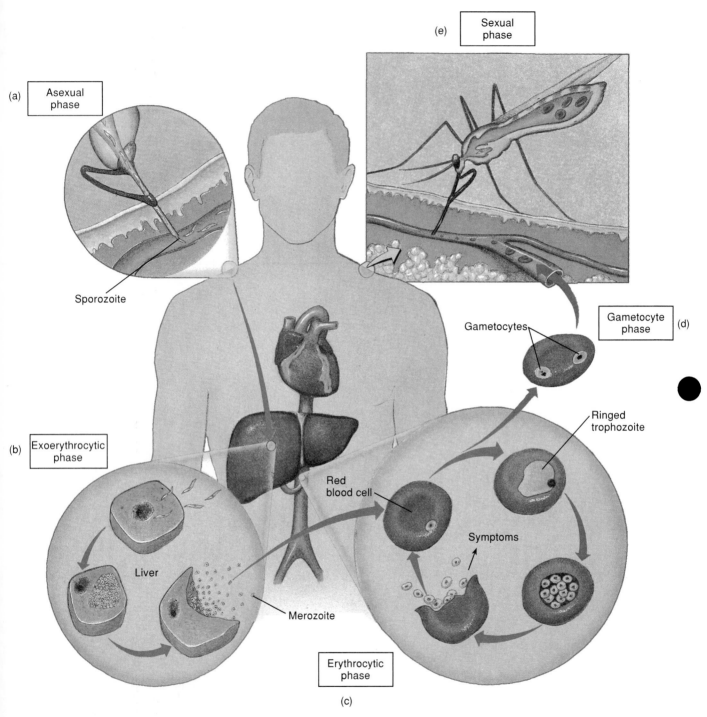

(e) Sexual phase

(a) Asexual phase

Sporozoite

Exoerythrocytic phase (b)

Liver

Merozoite

Gametocytes

Gametocyte phase (d)

Ringed trophozoite

Red blood cell

Symptoms

Erythrocytic phase

(c)

Life Cycle and Transmission of *Plasmodium*
Figure 23.11

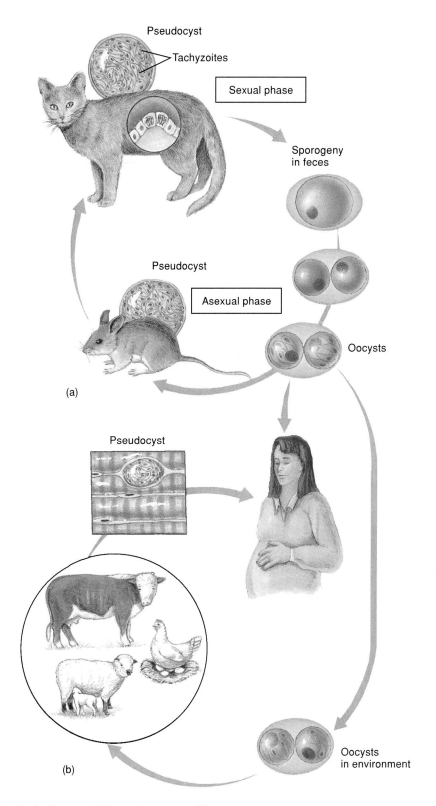

Pseudocyst

Tachyzoites

Sexual phase

Sporogeny
in feces

Pseudocyst

Asexual phase

Oocysts

(a)

Pseudocyst

(b)

Oocysts
in environment

Life Cycle and Morphologic Forms of *Toxoplasma gondii*
Figure 23.13

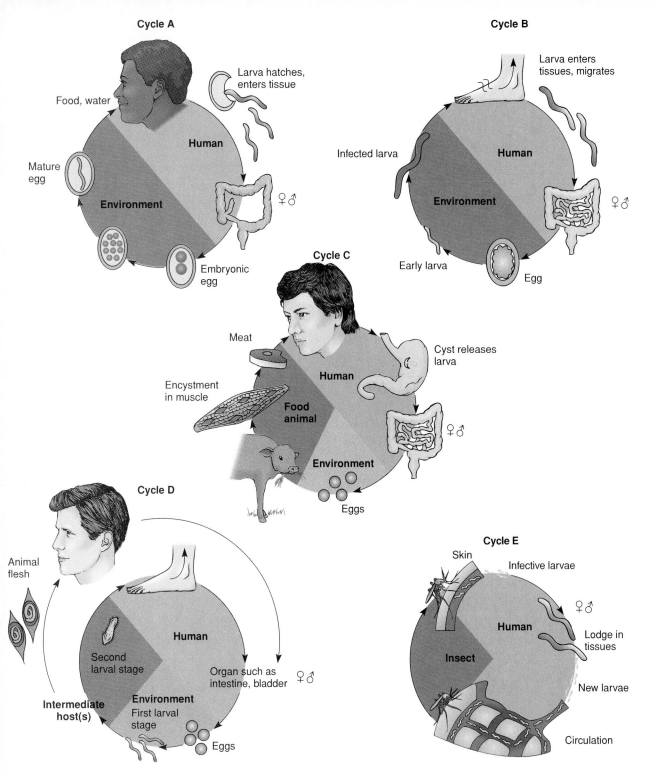

Cycle A

Food, water

Larva hatches,
enters tissue

Human

Mature
egg

Environment

Embryonic
egg

Cycle B

Larva enters
tissues, migrates

Infected larva

Human

Environment

Early larva

Egg

Cycle C

Meat

Cyst releases
larva

Human

Encystment
in muscle

Food
animal

Environment

Eggs

Cycle D

Animal
flesh

Human

Second
larval stage

Organ such as
intestine, bladder

Intermediate
host(s)

Environment
First larval
stage

Eggs

Cycle E

Skin

Infective larvae

Human

Insect

Lodge in
tissues

New larvae

Circulation

Five Basic Helminth Life and Transmission Cycles
Figure 23.16

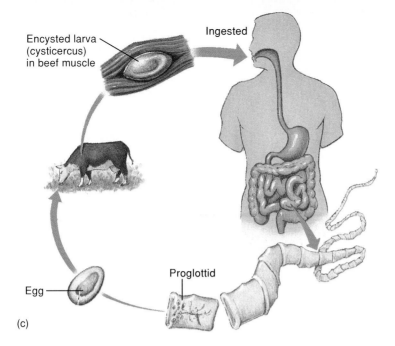

Encysted larva
(cysticercus)
in beef muscle

Ingested

Proglottid

Egg

(c)

The Life Cycle of *Taenia saginata*
Figure 23.26c

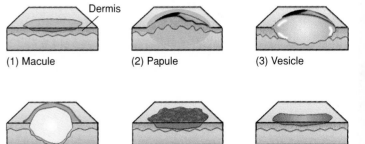

Dermis

(1) Macule

(2) Papule

(3) Vesicle

(4) Pustule

(5) Crust

(6) Scar

Stages in Pock Development
Figure 24.1

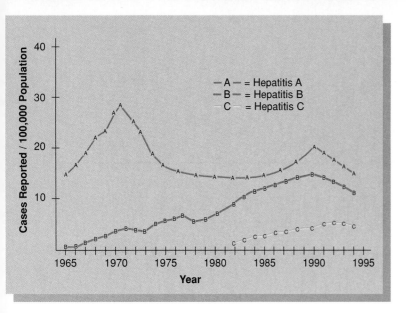

Comparative Incidence of Viral Hepatitis in the U.S., 1965-1995
Figure 24.16

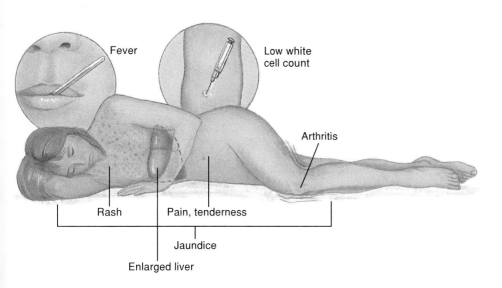

The Clinical Features of Hepatitis B
Figure 24.18

Enveloped	Nonenveloped

Segmented, Single-Stranded, Negative-Sense Genome*
    Orthomyxoviridae: Influenza
    Bunyaviridae: California encephalitis virus
                 Hantavirus hemorrhagic fever
    Arenaviridae: Hemorrhagic fevers
              Lassa fever virus
              Argentine hemorrhagic
                fever virus

Nonsegmented, Single-Stranded, Negative Sense

    Paramyxoviridae: Mumps virus
                 Measles virus
                 Respiratory syncytial virus

    Rhabdoviridae: Rabies virus
              Vesicular stomatitis virus

    Filoviridae: Ebola fever virus
            Marburg virus

Nonsegmented, Single-Stranded, Positive Sense
    Togaviridae: Rubella virus
             Western and Eastern equine
              encephalitis
             Dengue fever virus
             Yellow fever virus

    Coronaviridae: Common cold virus

Single-Stranded, Positive Sense, Reverse Transcriptase
    Retroviridae: AIDS (HIV)
            T-cell leukemia virus
            Hairy cell leukemia virus

Nonsegmented, Single-Stranded, Positive-Sense Genome
    Picornaviridae: Polio virus
              Hepatitis A virus
              Rhinoviruses

    Caliciviridae: Norwalk agent

Segmented, Double-Stranded, Positive Sense, Double Capsid
    Reoviridae: Tick fever virus
           Rotavirus gastroenteritis

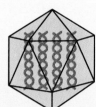

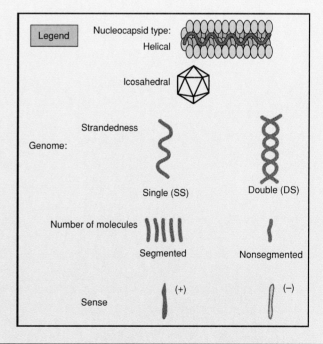

*If the RNA of the virus is in a form ready to be translated by the host's machinery, it is considered a positive-sense genome, and if it is not directly translatable by the host, it is a negative-sense genome.

RNA Virus Families
Table 25.1

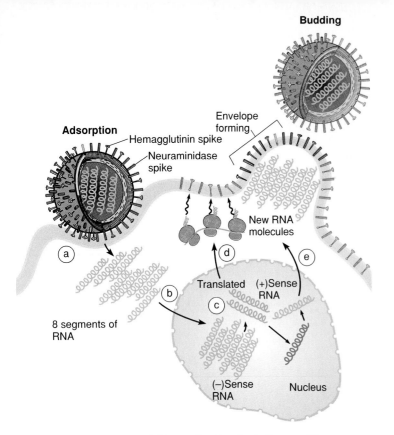

Adsorption

Hemagglutinin spike

Neuraminidase spike

Budding

Envelope forming

(a)

(b)

(c)

(d)

(e)

8 segments of RNA

Translated

New RNA molecules

(+)Sense RNA

(−)Sense RNA

Nucleus

Stages in Cell Invasion and Disruption by the Influenza Virus
Figure 25.1

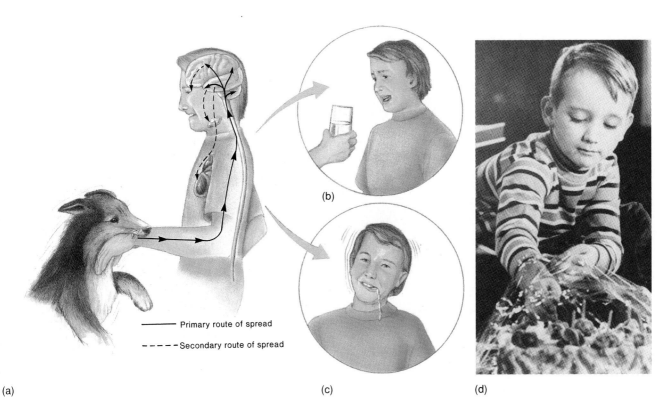

(a)

(b)

(c)

(d)

——— Primary route of spread

- - - - - Secondary route of spread

A Pathological Picture of Rabies
Figure 25.7

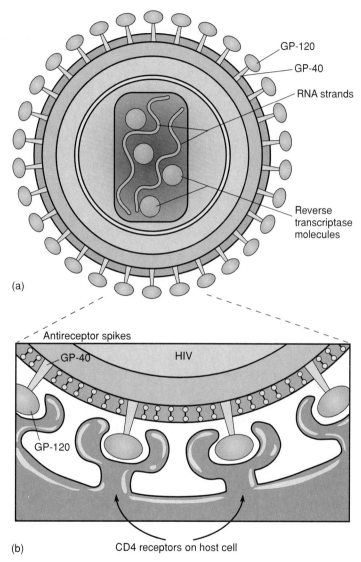

GP-120

GP-40

RNA strands

Reverse
transcriptase
molecules

(a)

Antireceptor spikes

GP-40

HIV

GP-120

CD4 receptors on host cell

(b)

A Cutaway Model of HIV
Figure 25.12

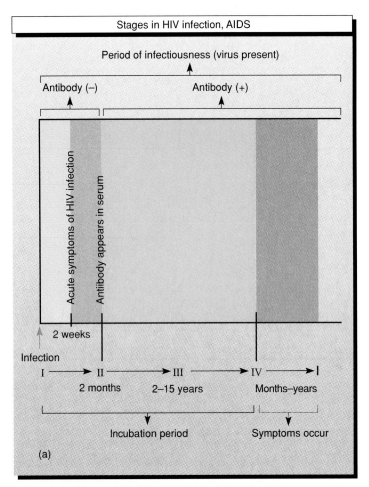

(a)

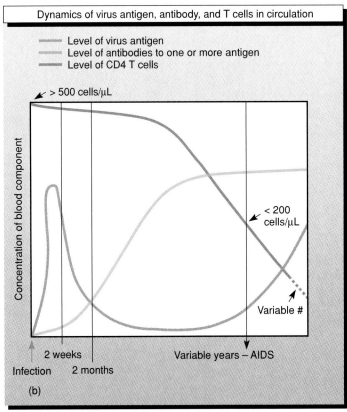

(b)

Stages of AIDS
Figure 25.15

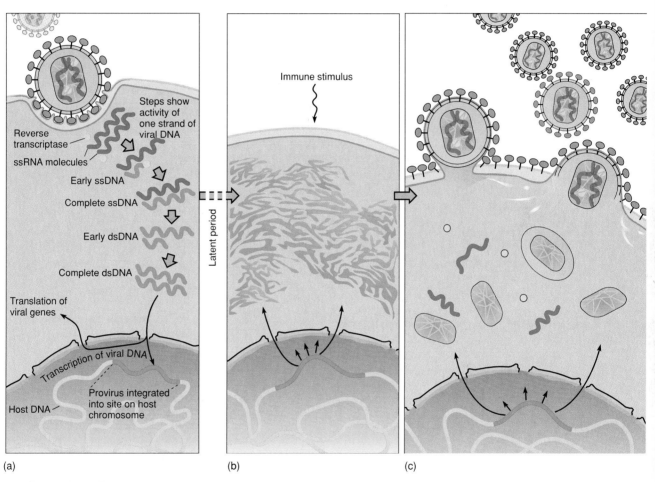

Reverse transcriptase
ssRNA molecules
Steps show activity of one strand of viral DNA
Early ssDNA
Complete ssDNA
Early dsDNA
Complete dsDNA
Translation of viral genes
Transcription of viral DNA
Host DNA
Provirus integrated into site on host chromosome

(a)

Latent period

Immune stimulus

(b)

(c)

The General Life Cycle of HIV
Figure 25.16

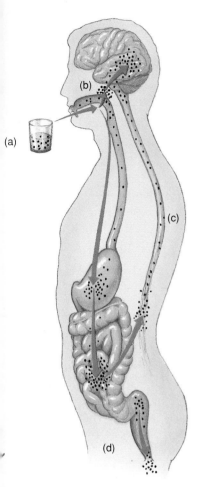

(a)

(b)

(c)

(d)

Stages of Infection and Pathogenesis of Poliomyelitis
Figure 25.25

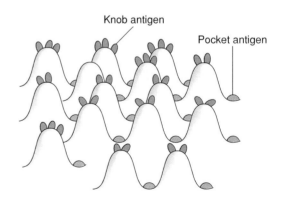

Knob antigen

Pocket antigen

(a)

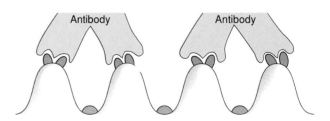

Antibody

Antibody

(b)

Surface of Rhinovirus
Figure 25. 28

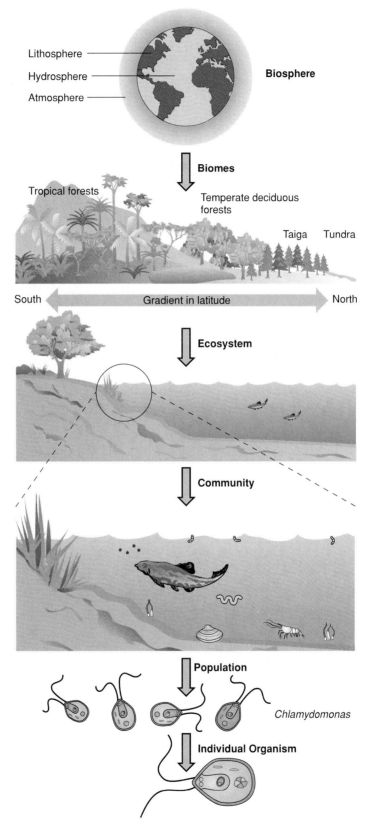

Lithosphere

Hydrosphere

Atmosphere

Biosphere

Biomes

Tropical forests

Temperate deciduous
forests

Taiga   Tundra

South   Gradient in latitude   North

Ecosystem

Community

Population

Chlamydomonas

Individual Organism

Levels of Organization in an Ecosystem
Figure 26.1

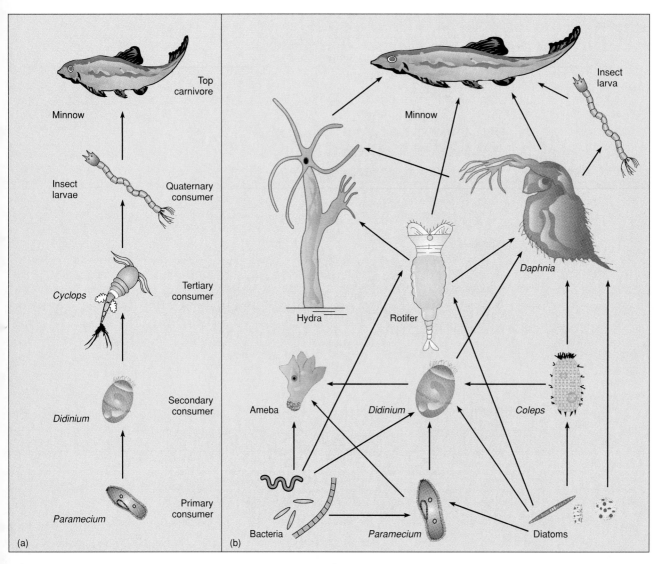

Comparison of a Food Chain and a Food Web in an Aquatic Ecosystem
Figure 26.3

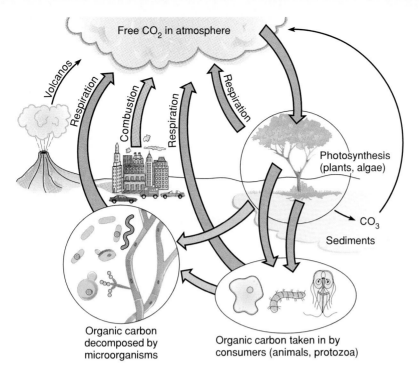

The Carbon Cycle
Figure 26.6

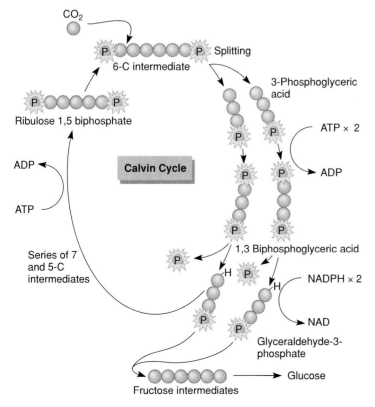

The Calvin Cycle
Figure 26.9

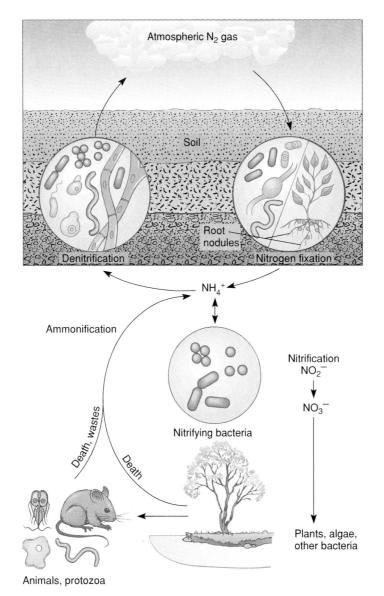

Atmospheric N$_2$ gas

Soil

Root nodules

Denitrification

Nitrogen fixation

NH$_4^+$

Ammonification

Nitrifying bacteria

Nitrification
NO$_2^-$

NO$_3^-$

Death, wastes

Death

Plants, algae, other bacteria

Animals, protozoa

The Nitrogen Cycle
Figure 26.10

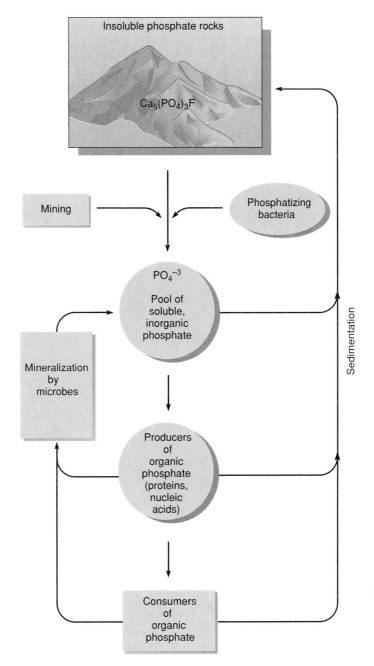

Insoluble phosphate rocks

$Ca_5(PO_4)_3F$

Mining

Phosphatizing bacteria

PO_4^{-3}

Pool of soluble, inorganic phosphate

Mineralization by microbes

Producers of organic phosphate (proteins, nucleic acids)

Consumers of organic phosphate

Sedimentation

The Phosphorus Cycle
Figure 26.13

153

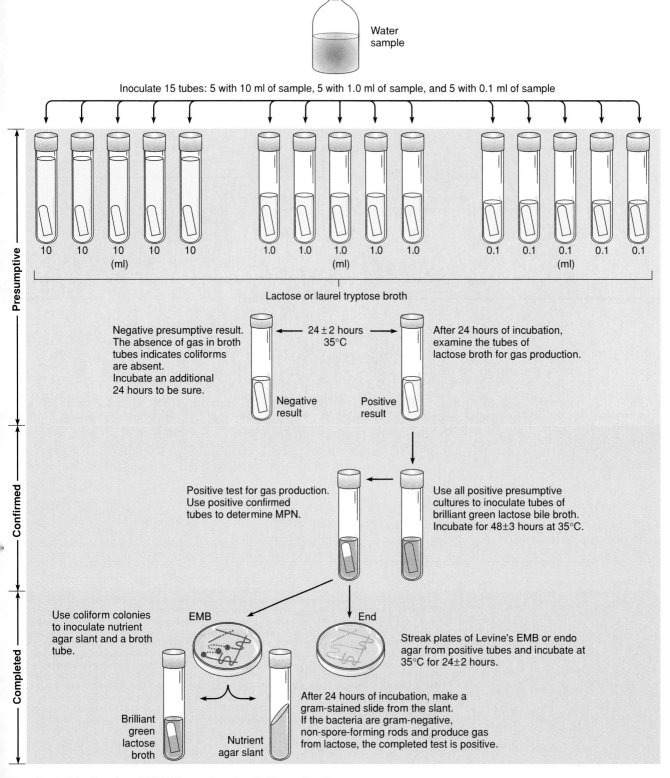

Water sample

Inoculate 15 tubes: 5 with 10 ml of sample, 5 with 1.0 ml of sample, and 5 with 0.1 ml of sample

Presumptive

| 10 | 10 | 10 | 10 | 10 | 1.0 | 1.0 | 1.0 | 1.0 | 1.0 | 0.1 | 0.1 | 0.1 | 0.1 | 0.1 |

(ml)   (ml)   (ml)

Lactose or laurel tryptose broth

Negative presumptive result. The absence of gas in broth tubes indicates coliforms are absent. Incubate an additional 24 hours to be sure.

Negative result

◄— 24 ± 2 hours 35°C —►

Positive result

After 24 hours of incubation, examine the tubes of lactose broth for gas production.

Positive test for gas production. Use positive confirmed tubes to determine MPN.

Use all positive presumptive cultures to inoculate tubes of brilliant green lactose bile broth. Incubate for 48±3 hours at 35°C.

Confirmed

Use coliform colonies to inoculate nutrient agar slant and a broth tube.

EMB

End

Streak plates of Levine's EMB or endo agar from positive tubes and incubate at 35°C for 24±2 hours.

Completed

Brilliant green lactose broth

Nutrient agar slant

After 24 hours of incubation, make a gram-stained slide from the slant. If the bacteria are gram-negative, non-spore-forming rods and produce gas from lactose, the completed test is positive.

Most Probable Number (MPN) Procedure for Coliform Testing
Figure 26.22

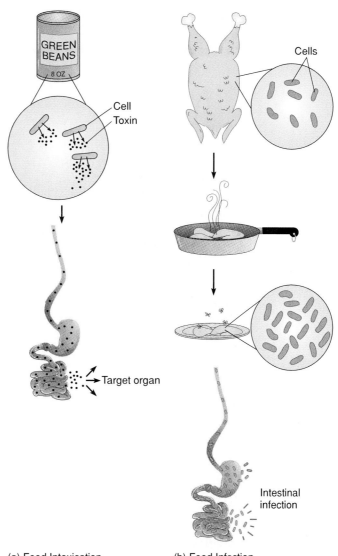

(a) Food Intoxication   (b) Food Infection

Food-Borne Illnesses of Microbial Origin
Figure 26.27

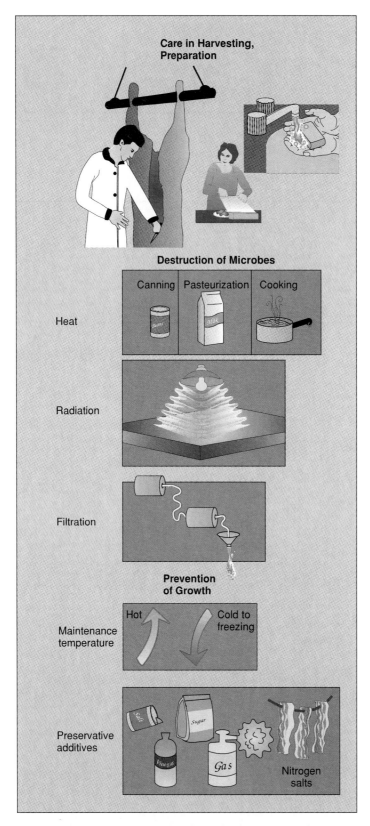

Primary Methods of Preventing Food Poisoning and Spoilage
Figure 26.28

CREDITS

Line Art

Fig. 1.3a: Data from World Health Organization 1994 reports.

Fig. 4.7: From Lansing M. Prescott, et al., *Microbiology,* 2nd edition. Copyright © 1993 Wm. C. Brown Communications, Inc. Reprinted by permission Times Mirror Higher Education Group, Inc., Dubuque, Iowa. All Rights Reserved.

Fig. 4.36: From Lansing M. Prescott, et al., *Microbiology,* 2nd edition. Copyright © 1993 Wm. C. Brown Communications, Inc. Reprinted by permission Times Mirror Higher Education Group, Inc., Dubuque, Iowa. All Rights Reserved.

Fig. 14.3a: From Kent M. Van De Graaff, *Human Anatomy,* 3rd edition. Copyright © 1992 Wm. C. Brown Communications, Inc. Reprinted by permission of Times Mirror Higher Education Group, Inc., Dubuque, Iowa. All Rights Reserved.

Fig. 14.9: From John W. Hole, Jr., *Human Anatomy and Physiology,* 6th edition. Copyright © 1993 Wm. C. Brown Communications, Inc. Reprinted by permission of Times Mirror Higher Education Group Inc., Dubuque, Iowa. All Rights Reserved.

Fig. 15.18: From Joseph A. Bellanti, *Immunology III.* Copyright © 1985, W.B. Saunders and Co., Philadelphia, PA. Reprinted by permission.

Fig. 16.6: From Lansing M. Prescott, et al., *Microbiology,* 2nd edition. Copyright © 1993 Wm. C. Brown Communications, Inc. Reprinted by permission Times Mirror Higher Education Group, Inc.,Dubuque, Iowa. All Rights Reserved.

Fig. 16.12: From Lansing M. Prescott, *Microbiology,* 2nd edition. Copyright © 1993 Wm. C. Brown Communications, Inc. Reprinted by permission of Times Mirror Higher Education Group, Inc., Dubuque, Iowa. All Rights Reserved.

Fig. 24.16: Source: Data from the Centers for Disease Control and Prevention, Atlanta, GA.

Fig. 26.22: From Lansing M. Prescott, *Microbiology,* 2nd edition. Copyright © 1993 Wm. C. Brown Communications, Inc. Reprinted by permission of Times Mirror Higher Education Group, Inc., Dubuque, Iowa. All Rights Reserved.

Photographs

Fig. 2.22d: From A.S. Moffat "Nitrogenase Structure Revealed," *Science*250:1513, December 14, 1990. © 1990 by the AAAS. Photo by M. M. Georgiadis and D.C. Rees, Caltech

Fig. 3.4b,d: © Kathy Talaro/Visuals Unlimited

Fig. 4.15a2,b2: © T.J. Beveridge/Visuals Unlimited

Fig. 6.24a1: © Carroll H. Weiss/Camera M.D. Studios

Fig. 6.24b1: © A.M. Siegelman/Visuals Unlimited

Fig. 6.24b2: © Science VU/CDC/Visuals Unlimited

Fig. 6.24c1: Fred P. Williams, Jr., U.S. Environmental Protection Agency

Fig. 6.24c2: National Institute of Health

Fig. 6.24d1: Courtesy of Diamedix

Fig. 6.24d2: Courtesy of Reference Laboratory, A PCC Laboratory

Fig. 14.24b: Reprinted from Wendell F. Rosse et al, "Immune Lysis of Normal Human and Paroxysmal Nocturnal Hemogloginuria (PNH) Red Blood Cells", *Journal of Experimental Midicine,* 123:969, 1966. Rockefeller University Press.

Fig. 15.14b: © R. Feldmann/Rainbow

Fig. 22.7b: © A.M. Siegelman/Visuals Unlimited

Fig. 22.8b: © Science VU-Charles Sutton/Visuals Unlimited

Fig. 25.7b: © AP/Wide World Photos